Praise for *The Grand Bioc*

"For those addicted to exploring our role as observers in defining our universe, here is your long-awaited major update. It is as rare as a unicorn to see a major stem cell biologist collaborating with a theoretical physicist to produce a coherent, fresh-new, readable, and clearly illustrated book. You'll love *The Grand Biocentric Design*—it adds new turf to the physics of making universes, and includes 'solid evidence,' at last, that observers define the structure of physical reality itself."

—George Church, Robert Winthrop Professor of Genetics at Harvard Medical School, professor of health sciences and technology at Harvard and MIT, and a member of the National Academy of Sciences and the National Academy of Engineering (on Thomson Reuters short-list for the Nobel Prize)

"*The Grand Biocentric Design* brilliantly draws our attention to the most important feature of the entire universe: our human minds. Robert Lanza deeply appreciates, and eloquently analyzes, the penetration that our human minds have achieved of the underlying mathematical machinery of that universe, exposing its deeply and purely mental character. In fact, it is the physicists' microscopic examination of that external world that most vividly reveals the 'the grand biocentric design' of the universe. This new book brings out the real nature of our universe: for all of us to deeply search for fuller understanding, and for meaning."

—Richard Conn Henry, academy professor of physics and astronomy at The Johns Hopkins University, and former deputy director of NASA's Astrophysics Division

"In 1887, the Michelson and Morley experiment unexpectedly showed that the celestial aether did not exist. This upended classical physics, led to Einstein's theory of relativity and the atomic bomb. Since then, quantum physics has revealed the decisive role that the observer's consciousness plays in shaping the world we experience. These experiments have also upended our assumption that time and space have objective reality. As before, a new theory waits to be born. Biocentrism is such a theory . . . For those searching for answers to contemporary physics' disturbing findings, *The Grand Biocentric Design* is a must-read."

—Ronald M. Green, Eunice and Julian Cohen Professor Emeritus for the study of ethics and human values at Dartmouth College, and Professor Emeritus and former chairman of the Department of Religion.

"Even as a child, Robert Lanza believed that living things were the subjects most worthy of scientific study. Now, in *The Grand Biocentric Design*, his third and best book on the topic, Lanza and colleagues unpack, with unprecedented rigor, his theory of biocentrism through the hard lens of physics. It takes the oddities of quantum physics to a new level, defining reality itself and giving ballast to the seductive idea that time travel is possible, death an illusion, and life, a perpetual flower in bloom. If you consider biocentrism mere philosophy, look to this volume to make the case that science is at its core."

—Pamela Weintraub, senior editor at *Aeon*, former executive editor of *Discover*, and editor-in-chief of *OMNI*

"In his two previous books on biocentrism (written with Bob Berman), biologist Robert Lanza proposed a bold new theory of the universe, one that builds on the insights of quantum physics to put consciousness at its center. Here, with theoretical physicist Matej Pavšič, Lanza strives, in language suited to the general reader, to explain the science behind this theory. Its stark differentness from the materialist view of the universe makes the mind rebel, but thinkers as various as Ralph Waldo Emerson and Stephen Hawking have had intimations of what these writers describe."

—Robert Wilson, editor in chief at *The American Scholar*, the venerable magazine of Phi Beta Kappa, which has published the work of Albert Einstein, John Updike, Saul Bellow, Bertrand Russell, Margaret Mead, and Robert Frost, among others

"Once again, Lanza and colleagues continue to guide readers who have a quest to understand our universe. This must-read book is a masterpiece, discussing newly emergent research that answers questions, through the lens of biocentrism, on how the world works and who we are. If you've ever stood on the beach at night looking up at the vast sky with thoughts of how and why, then the breakthroughs presented about reality and consciousness, and the experience of time and how we perceive it, will provide thought provoking and life changing insights on your existence and everything that surrounds you."

"Robert Lanza is one of the most creative and brilliant scientists I have ever known. Ever since he became convinced that living things are the subjects worthy of scientific study when he was very young, he has dedicated his career to biology and life. *The Grand Biocentric Design* is his latest creative work based on his life-long scientific journey, which opens up a new biology-based vista to our understanding of existence and consciousness."

"The revolution in quantum mechanics introduced perplexing counterintuitive ideas that were outside the realm of human experience, including wave-particle duality, quantization of molecular structure, Schrodinger's cat, and the Heisenberg uncertainty principle, to name a few. Important paradoxes remain unexplained including quantum coupling between particles separated by great distances (i.e., action at a distance idea). Into this conundrum comes *The Grand Biocentric Design* by Robert Lanza, with theoretical physicist Matej Pavšič, with a unique and paradigm-shattering concept that biological systems are primary and affect our perception of physical systems. Lanza is an accomplished stem cell biologist and original thinker who expands his ideas on the interplay between biology and physics in this fabulous book that is approachable by an educated lay audience. This insightful work is certain to energize our conversations about the nature of the biological and physical world."

"It's fabulous—I couldn't put it down! A masterly tour de force that will change your life. Robert Lanza and his coauthors take on the Herculean task of reconciling quantum theory, relativity, and consciousness. You will never look at science—indeed, life and death—the same way again."

THE GRAND
BIOCENTRIC
DESIGN

Also by Robert Lanza and Bob Berman
Biocentrism
Beyond Biocentrism

Also by Matej Pavšič
The Landscape of Theoretical Physics: A Global View

THE **GRAND BIOCENTRIC DESIGN**

How Life Creates Reality

ROBERT LANZA, MD
and MATEJ PAVŠIČ

with BOB BERMAN

BenBella Books, Inc.
Dallas, TX

BenBella Books, Inc.
10440 N. Central Expressway
Suite 800
Dallas, TX 75231
www.benbellabooks.com
Send feedback to feedback@benbellabooks.com

BenBella is a federally registered trademark.

Printed in the United States of America
10 9 8 7 6 5 4 3 2 1

ISBN 9781953295804 (trade paperback)

The Library of Congress cataloged the hardcover as follows:
Names: Lanza, R. P. (Robert Paul), 1956- author. | Pavšič, Matej, author. | Berman, Bob, author.
Title: The grand biocentric design : how life creates reality / Robert Lanza, MD and Matej Pavšič with Bob Berman.
Description: Dallas, TX : BenBella Books, Inc., [2020] | Includes bibliographical references and index. | Summary: "A new installment in the series that blew readers' minds with Biocentrism and Beyond Biocentrism, The Grand Biocentric Design offers an even deeper dive in to the nature of reality and our universe based on the latest groundbreaking research"—Provided by publisher.
Identifiers: LCCN 2020028855 (print) | LCCN 2020028856 (ebook) | ISBN 9781950665402 (hardback) | ISBN 9781950665556 (ebook)
Subjects: LCSH: Time and space—Philosophy. | Time—Philosophy. | Cosmology—Philosophy. | Metaphysics.
Classification: LCC QC173.59.S65 L335 2020 (print) | LCC QC173.59.S65 (ebook) | DDC 523.1—dc23
LC record available at https://lccn.loc.gov/2020028855
LC ebook record available at https://lccn.loc.gov/2020028856

Editing by Alexa Stevenson
Copyediting by Scott Calamar
Proofreading by Michael Fedison and Sarah Vostok
Indexing by WordCo Indexing Services
Text design and composition by Aaron Edmiston
Cover design by Sarah Avinger
Cover photo © Shutterstock / betibup33
Interior illustrations by Jacqueline Rogers
Printed by Lake Book Manufacturing

Special discounts for bulk sales are available.
Please contact bulkorders@benbellabooks.com.

To Eliot Stellar—The man who cared
(see Post Scriptum for story)

University Archives and Records Center, University of Pennsylvania

Eliot Stellar (1919–1993)

One of the founders of behavioral neuroscience—shown here at his desk in 1978 when he was Lanza's advisor.

"Stellar dedicated much of his time in his later years to the Human Rights Committee of the National Academy of Sciences (NAS), serving as its Chairman from 1983 until the end of his life. In his work for the NAS he actively lobbied for the freedom of scientists to conduct their work throughout the world and interceded in behalf of imprisoned scientists who were in danger of losing their lives or suffering great hardships."

—From the "Eliot Stellar Papers,"
University of Pennsylvania Archives

CONTENTS

Copernicus dethroned humanity from the cosmic center.
Does quantum theory suggest that, in some mysterious
sense, we are a cosmic center?
— **Bruce Rosenblum and Fred Kuttner,**
Quantum Enigma

INTRODUCTION

ROBERT LANZA

In all directions, the current scientific paradigm leads to insoluble enigmas, to conclusions that are ultimately irrational. Since World Wars I and II there has been an unprecedented burst of discovery, with findings that suggest the need for a fundamental shift in the way science views the world. When our worldview catches up with the facts, the old paradigm will be replaced with a new *bio*centric model, in which life is not a product of the universe, but the other way around.

A change to our most foundational of beliefs is bound to face resistance. I'm no stranger to this; I've encountered opposition to new ways of thinking my whole life. As a boy, I lay awake at night and imagined my life as a scientist, peering at wonders through a microscope. But reality seemed determined to remind me that this was only a dream. Upon entering first grade, students at my elementary school were separated into three classes based upon their perceived "potential"—A, B, and C. Our family had just moved to the suburbs from Roxbury, one of the roughest areas of Boston (it was later razed for urban renewal). My father was a professional gambler (he played cards for a living, which at the time was illegal—not to mention the dog and horse tracks), and our family was not exactly considered scholarly material. Indeed, all three of my sisters subsequently dropped out of high school. I was placed in the

C class, a repository for those destined for manual, trade labor, a class that included the students who had been kept back and those who were mainly known for shooting spitballs at teachers.

My best friend was in the A class. "Do you think I could become a scientist?" I asked his mother one day in fifth grade. "If I tried hard, could I be a doctor?"

"Good gracious!" she responded, explaining that she'd never known anyone in the C class to become a doctor, but that I'd make an excellent carpenter or plumber.

The next day I decided to enter the science fair, which put me in direct competition with the A class. For his project on rocks, my best friend's parents took him to museums for his research and created an impressive display for his specimens. My project—animals—was made up of souvenirs from my various excursions: insects, feathers, and bird eggs. Even then I was convinced that living things—not inert material and rocks—were the subjects most worthy of scientific study. This was a complete reversal of the hierarchy taught in our schoolbooks—that is, the realm of physics, with its forces and atoms, forming the foundation of the world and thus most key to its understanding, followed by chemistry and then biology and life. My project won me, a lowly member of the C class, second place behind my best friend.

Science fairs became a way to show up those who labeled me for my family's circumstances. By trying earnestly, I believed I could improve my situation. In high school, I applied myself to an ambitious attempt to alter the genetic makeup of white chickens and make them black using nucleoprotein. It was before the era of genetic engineering, and my biology teacher said it was impossible; my chemistry teacher was blunter, saying, "Lanza, you're going to hell."

Before the fair, a friend predicted I'd win. "Ha-ha!" the whole class laughed. But my friend was right.

Once, after my sister was suspended, the principal had told my mother that she wasn't fit to be a parent. When I won, that principal had to congratulate my mother in front of the whole school.

I did go on to become a scientist, and during my scientific career, I continued to encounter intolerance to new ideas. *Can you generate stem*

This is the certificate the author (Lanza) received for his C class science project on "Animals." It was cosigned by Barbara O'Donnell—his future junior high school science teacher—who nurtured his scientific growth, as she did for hundreds of other students during her fifty years as a teacher and guidance counselor. The book Biocentrism was dedicated to her on the occasion of her ninetieth year.

cells without destroying embryos? Can you clone one species using eggs from another? Could findings at the subatomic level "scale up" to tell us something about life and consciousness? Scientists are trained to ask questions, but they are also trained to be cautious and rational; their questioning is often aimed at the incremental change, not the paradigm-toppling one. After all, scientists are no different from the rest of our species. We evolved in the forest roof to collect fruit and berries while evading predators and staying alive long enough to procreate; it shouldn't come as any surprise that this skill set hasn't always served us perfectly in understanding the nature of existence.

"One thing I have learned in a long life," said Einstein, "[is] that all our science, measured against reality, is primitive and childlike—and yet it is the most precious thing we have." Science must work with simple concepts the human mind can comprehend. But as the evidence for biocentrism mounts, science may prove the key to answering questions previously thought to be beyond its borders, those that have plagued us since before the beginning of civilization.

* * *

This may be the beginning of this book, but it is not the beginning of our story.

That's because we are plunging into an ongoing odyssey. It's a movie that has already started, and we are seating ourselves long after the opening credits have rolled.

As we will soon see, the Renaissance witnessed a transformation in the way humans attempted to understand the cosmos. But even as superstition and fear slowly lost their grip, the established view that emerged dictated a firm division between two basic entities—we observers glued to the surface of our small planet, and the vast realm of nature that constitutes a cosmos almost wholly separate from ourselves. The assumption that these entities are two entirely different balls of wax has so permeated scientific thought that it is likely still assumed by the reader even now in the twenty-first century.

However, the opposing view is hardly new. Early Sanskrit and Taoist teachers unanimously declared that when it comes to the cosmos, "All is One." Eastern mystics and philosophers inherently perceived or intuited a unity between the observer and the so-called external universe, and, as centuries elapsed, they were consistent in maintaining that such a distinction is illusory. Some Western philosophers, too—among them Berkeley and Spinoza—challenged the prevailing views about the existence of an external world and its separation from consciousness. Nonetheless, the dichotomous paradigm remained the majority consensus, especially in the world of science.

But the maverick minority got a major megaphone a century ago, when some of the originators of quantum theory—most notably Erwin Schrödinger and Niels Bohr—concluded that consciousness is central to any true understanding of reality. While they reached their conclusions by way of advanced math, in the course of developing the equations that would form the basis for quantum mechanics and its innumerable successes, they thus were also pioneers who helped set the table for biocentrism a century later.

Today, oddities of the quantum world like entanglement have moved the minority increasingly into the mainstream. If it's really true that life and consciousness are central to everything else, then countless puzzling anomalies in science enjoy immediate clarification. It's not just bizarre laboratory results like the famous "double-slit experiment" that make no sense unless the observer's presence is intimately intertwined with the results. On an everyday level, hundreds of physical constants such as the strength of gravity and the electromagnetic force called "alpha" that governs the electrical bonds in every atom are identical throughout the universe and "set in stone" at precisely the values that allow life to exist. This could merely be an astounding coincidence. But the simplest explanation is that the laws and conditions of the universe allow for the observer because the observer generates them. Duh!

This is also a story in progress because we've told some of it in two previous books on biocentrism—many of you may have already read one or both of these. If so, you won't be faulted for wondering why this third book was necessary. The short answer is that this book both outlines biocentrism in a new way and also expands upon it.

In the first two biocentrism books, *Biocentrism* and *Beyond Biocentrism*, we employed a wide spectrum of tools to show why everything makes far more sense if nature and the observer are actually intertwined, or correlative—using not just science but also basic logic and the assessments of some of the great thinkers through the centuries. Our multipronged approach to explaining and reinforcing our conclusions has been both persuasive and popular, as demonstrated by the great success of those first biocentrism books, which have been translated into two dozen languages, with editions published around the world. And yet some science-minded readers wanted more.

To some of those readers, biocentrism's conclusions about consciousness skirted the category of "woo," meaning scientifically dubious, New Agey–type theorizing. Such comments gave us pause. Might our hard-won conclusions, though fundamentally based on cold logic and hard science, still amount to a mere "philosophical" interpretation of the experimental and observational results? Did biocentrism more properly fall under the rubric of philosophy than of science? We certainly didn't think so. Yet we acknowledged that it would be nice to be able to seal the case for biocentrism on the physics alone.

What's more, since the first two books were released, new research has emerged that makes the case for biocentrism stronger than ever, allowing us to explain formerly fuzzy aspects of how our biocentric universe actually works. As our understanding has grown, we've been able to refine our theory and build upon it, discovering new core principles that demand inclusion in any complete accounting of biocentrism. It was time for a newly comprehensive view of the grand biocentric design governing our cosmos.

That's what's in front of you now. As you'll see, this present volume tells our story in a way that relies solely on the hard sciences. We've confined the equations and such to the appendixes, since we know that many readers will slam a book shut at the mere sight of a square-root symbol. Because while rigorously scientific, we want this to be a fun exploration for the general public, too—after all, the questions this book answers are those every one of us has asked, basic questions about life and death, about how the world works and why we exist.

What follows is not an exhaustive treatment since we've omitted lengthy discussion of some things, like the double-slit experiment, that were covered fully in the previous books. Nevertheless, we will recount the history of astounding physics discoveries that all lead inexorably to the bizarre but reality-shaking conclusion that the basic structure of the cosmos—things like space and time and the way matter holds together—requires observers. Though many physicists define the observer as any macroscopic object, we are among those who believe the observer must be a conscious one. More about why—and what that means—later on.

As our story unfolds, we will see how Newton's laws not only determined how things actually move, but also how an object could have moved if it started out another way, bringing with them the first faint breezes of alternate universes and foreshadowing quantum theory.

We'll visit the rise of that theory, and the discovery of the strange quantum behavior that challenged the idea that an external world exists independent of the perceiving subject—an idea debated by philosophers and physicists from Plato to Hawking. We'll dive into what Niels Bohr, the great Nobel physicist, meant when he said, "We're not measuring the world; we're creating it."

We'll untangle the logic that the mind uses to generate our spatiotemporal experience, and get insights into the so-called "hard problem" of how consciousness arises, exploring those quantumly entangled regions of the brain that together constitute the system we associate with the unitary "me" feeling. We explain, for the first time ever, the entire mechanism involved in the emergence of what we experience as time—from the quantum level, where everything is still in superposition, to the macroscopic events occurring in the brain's neurocircuitry. Along the way, we'll see how information that breaks the light-speed limit suggests the mind is unified with matter and the world.

As we increasingly recognize life as an adventure that transcends our commonsense understanding, we will also get hints about death. We'll look at the mind-twisting thought experiment called quantum suicide, which can be used to explain why we are here now despite the overwhelming odds against it—and why death has no true reality. We will see that life has a nonlinear dimensionality, like a perennial flower that always blooms.

Throughout the book, we will find countless commonsense assumptions turned on their heads. For instance: "The histories of the universe," said the late theoretical physicist Stephen Hawking, "depend on what is being measured, contrary to the usual idea that the universe has an objective observer-independent history." While in classical physics the past is assumed to exist as an unalterable series of events, quantum physics plays by a different set of rules in which, as Hawking said, "the past, like the future, is indefinite and exists as a spectrum of possibilities."

And while we're at it, we'll look at physicists' century-long frustration at that very fact: that quantum mechanics exists via a "different set of rules." After all, making sense of gravity, among other things, requires finding a way to reconcile Einstein's theory of general relativity, which accurately describes the macroscopic, large-scale cosmos, with the altogether different rules governing the quantum realm of the tiny. Why can't science-at-large-scales communicate with science at the subatomic level? Astoundingly, this book arrives at a breakthrough in exactly that quest, a Holy Grail of physics.

That breakthrough comes in the final chapters, where we will encounter an astounding cover-story paper by one of the authors (Lanza) and Dmitriy Podolskiy, a theoretical physicist working at Harvard, that explains how time itself emerges directly from the observer. We will learn that time does not exist "out there," ticking away from past to future as we've always assumed, but rather it's an emergent property like a fast-growing bamboo stalk, and its existence depends on the observer's ability to preserve information about experienced events. In the world of biocentrism, a "brainless" observer does not only fail to experience time—without a conscious observer, time has no existence in any sense.

But this book is not merely an arrow targeted at the shocking revelations in the final chapters. Nor even at the full flabbergasting scientific evidence that there is simply no time, no reality, and no existence of any kind without an observer. Instead, it is an odyssey engineered to awe and inspire as it reveals the workings of the cosmos and our place in it.

So, yes, expect fireworks at the end, as the old paradigm is decisively replaced by the new. But watching this amazing story unfold is a journey that is its own reward, with surprises at every turn.

And it starts where we might least expect it, in the familiar if still-puzzling realm of simple everyday awareness.

"In the last analysis, we ourselves are part of the mystery we are trying to solve."

Max Planck
Nobel Prize 1918

"Consciousness cannot be accounted for in physical terms. For consciousness is absolutely fundamental."

Erwin Schrödinger
Nobel Prize 1933

"Contemporary science, today more than at any previous time, has been forced by Nature herself to pose again the question of the possibility of comprehending reality by mental processes."

Werner Heisenberg
Nobel Prize 1932

"We are not only observers, we are participators."

John Wheeler

"When we measure something, we are forcing an undetermined, undefined world to assume an experimental value. We are not 'measuring' the world, we are creating it."

Niels Bohr
Nobel Prize 1922

"The very study of the external world [leads] to the conclusion that the content of consciousness is an ultimate reality."

E.P. Wigner
Nobel Prize 1963

"There's no way to remove the observer–us–from our perceptions of the world ... the past, like the future, is indefinite and exists only as a spectrum of possibilities."

Stephen Hawking

Hints of the biocentric nature of the universe have been made by some of the greatest scientists in modern physics, including Planck, Schrödinger, Heisenberg, and Bohr–the founders of quantum mechanics.

FIGURING OUT THE UNIVERSE

1

All of us are prisoners of our early indoctrinations, for it is hard,
very nearly impossible, to shake off one's earliest training.
—**Jubal, in *Stranger in a Strange Land*, by Robert Heinlein**

These are perilous times for science. They are also exciting beyond comparison.

Perilous because in many countries, anti-science undercurrents threaten to dilute the amazing progress of the past few decades. Exciting because some of our deepest questions are at last being answered, and our most urgent human problems are on the cusp of being solved.

The changes wrought by scientific progress are most obvious when we compare the world today to that when some of us first studied science, in the midseventies. No space probe had ventured past Mars. No one knew that quarks formed the nucleus of every atom. The internet did not exist. Even VHS camcorders lay years in the future.

The average new car cost $3,700. A typical US house was $35,000.

In the intervening years, science has transformed the planet—from genetic engineering that now feeds a global population once thought to be unsustainable, to routine cardiac surgery and other advances that have pushed the average human life span into the eighties.

This book intends to push the boundaries of science even further.

As mentioned, we'll leave the head-scratching physics equations for the appendix, but we'll also assume you, the reader, enjoy an average degree of science knowledge. Maybe even above average: the National Science Foundation, which keeps track of public science awareness, recently released their annual basic knowledge survey, and the results weren't the sort anyone would want to display on the fridge.

The survey includes nine true/false questions—things like: 1. The center of the earth is very hot. 2. All radioactivity is man-made. 3. Electrons are smaller than atoms, and so on.* The public's performance on this test has not changed much in the past forty years—the average score works out to about 60 percent, a D. (And, contrary to what many believe, Europeans don't fare much better.)

Perhaps even more disturbing than the state of the public's knowledge is the state of critical thinking: Surveys reveal that a disturbing minority believes in various conspiracy theories. For example, polls show that 7 percent of the American public thinks the Apollo moon landings were a hoax. And in 2018, the fastest growing web conspiracy was that Earth is actually flat and that the photos supposedly taken of our planet from space have been faked. Sadly, such beliefs often persist in spite of being contradicted not by complicated or esoteric science but by basic common sense: In this case, a belief in a flat Earth can be disproved by a simple phone call between friends on the East and West Coasts of the US, since the sun appears halfway up the sky for those in California while, at that same moment, it is setting on the horizon for those in Vermont. This alone proves that our planet cannot be flat.

This book is not for those, like "flat-earthers," who refuse to believe the evidence before them. It is aimed instead at readers who are receptive to major revelations based on observations and experimentation—for

---

* In case high school science was a while ago for you, the answers are T, F, and T.

that is what biocentrism is, even if our ultimate focus centers on fundamental aspects of life that have previously appeared hopelessly mysterious and insoluble by science.

Having slogged through centuries of superstition that sometimes spurred brutal repressions of scientific progress (think Galileo), most of the modern world finally regards science as the surest source of knowledge about nature. As a bonus, it gives us our tech goodies—our iPhones and GPS. It offers us tomatoes in January.

Beyond that, the scientific method itself is the most effective truth-finding process ever devised. With its emphasis on skepticism, observation, and testing, it brutally hacks down pretenders. Anyone making an original far-out claim—as when Luis and Walter Alvarez claimed a meteor impact made the dinosaurs extinct—must come up with solid evidence. In the case of the Alvarez father-and-son team, this evidence was a worldwide deposition layer of iridium (rare on Earth but abundant in meteor dust) laid down 66 million years ago. Their resultant fame inspired other researchers to try to "shoot down" the Alvarez theory in order to gain their own renown and leave their own mark on history. Thus science provides ongoing motivation for offering antithetical views and skeptical analysis. It's self-regulating.

Unfortunately, as we discussed in the introduction, scientists are all too human, and science has its own inertia, which is why truly new ideas typically languish not merely for years but often for decades or even centuries. It is a sad fact that when German meteorologist Alfred Wegener came up with his theory of continental drift in 1912, it was widely dismissed even into the 1950s. Once finally accepted, it not only let everyone see the obvious—that continent boundaries fit like jigsaw puzzle pieces into each other, suggesting that they were once all part of a supercontinent we now call Pangaea—but also explained such oddities as mid-ocean seafloor spreading, and the fact that rocks in eastern North America closely resemble those in Ireland. It finally made sense of the "ring of fire" that borders the Pacific with frequent volcanic and seismic activity. In short, many mysteries were solved all at once by this new awareness of our planet's crust floating like flotsam upon molten magma and shifting by one to four inches a year—but this awareness took decades to dawn.

Other bits of epoxy that sometimes gum up the wheels of progress are nature's aspects that are so omnipresent that we're overaccustomed to them, which inhibits objective analysis. They're too common to cry out for attention.

Such familiarity may explain why air was not identified as being composed of discrete gases, each with very different characteristics, until after the American Revolution. You'll find no suggestion that air was anything but a single substance in the normally inquisitive writings of the ancient Greeks, or even those of early Renaissance geniuses.

This may be our situation today when it comes to consciousness. The fact that everything seen, heard, thought about, or remembered is, first and foremost, a manifestation of human awareness means that consciousness is so close and intimate it is usually overlooked. "Awareness" is like the screen upon which a movie is projected. It is the "thing that is real" as we sit in the theater, and yet we ignore it just as we don't see the flickering profusion of colors and lights that the projector has cast upon it as what they are. Instead, our focus remains trained upon the shapes the film creates, the patterns that we recognize as the faces of actors or the meanings conveyed by language encoded in the soundtrack.

But the cinema analogy only carries us so far. With a movie screen, there is no inherent importance in the curtain of reflective material; another surface such as a white wall would have sufficed. Consciousness is something else. The fact of awareness, of perception, is not only fundamental to all we know or ever hope to know, it is also exceedingly peculiar, both in fact and in origin.

Since knowledge is science's *sine qua non*, and perception is the only way to acquire knowledge, consciousness ought to seem more basic to our understanding than any neural methodology or subsystem. After all, if human consciousness contains fundamental biases or quirks, these could color everything we see and learn. So we'd want to know about this before proceeding further to our countless information acquisition methods, whether they be color and sound classifications or taxonomies of life forms. Consciousness is the root. It's more fundamental than your computer's hard drive. In this analogy, it's rather like the electric current.

Moreover, experiments since the 1920s have unequivocally revealed that the mere presence of the observer changes an observation. Treated then and now as an oddity or inconvenience, this phenomenon strongly suggests that we are not separate from the things we see, hear, and contemplate. Rather, we—nature and the observer—are some sort of inseparable entity. This simple conclusion lies at the heart of biocentrism.

But what is this entity? Unfortunately, since consciousness has only been studied superficially and largely remains a mystery, the amalgam that is "consciousness + nature" is equally enigmatic—in fact, more so. By "studied superficially," we mean that while neuroscience has progressed impressively from determining which parts of the brain control various sensory and motor functions to exploring how complicated networks of neurons encode concepts, this same field has done little to solve deep foundational problems such as how consciousness arises from matter in the first place—the so-called "hard problem of consciousness." Perhaps such researchers cannot be faulted, since those bedrock issues have proven stubbornly immune to elucidation via science's usual tools. How would *you* begin to design an experiment that results in objective information about this most subjective of phenomena?

Science has an established tradition when it comes to aspects of nature that defy logical explanation and resist experimentation. It ignores them. This is actually a proper response, since no one wants researchers offering fudgy guesses. Official silence may not be helpful, but it is respectable. However, as a result, the very word "consciousness" may seem out of place in science books or articles, despite the fact that, as we'll see, most famous names in quantum mechanics regarded it as central to the understanding of the cosmos. And that's before its somewhat newly acknowledged role not just in revealing what we observe, but creating it.

How human (and likely also nonhuman animal) awareness accomplishes such an unexpected but key function in nature is the primary focus of this book, and so we will explore consciousness in multiple chapters. This will include following the progress various disciplines have made in qualifying the act-of-observing process, and seeing how seemingly inanimate nature interacts with living awareness, which in turn is linked with complex neural architecture. One of the authors (Lanza, with the theoretical physicist Dmitriy Podolskiy) has recently made and published new discoveries about what actually unfolds at that critical consciousness/observation moment. As we'll see, it is a eureka revelation that—together with the other scientific findings discussed in the book—suggests the need for a Copernican-scale revolution.[†]

The public generally looks to science for aid or answers in three main categories, which have changed little over time. First and foremost, naturally, is the "what's in it for me?" category: people want science to give them cures for diseases, aids for defective vision or hearing, transportation improvements like reliable jetliners, and affordable personal gadgets such as cell phones. Their second tier of attention revolves around straightforward questions about the world—think new information about life on Mars, black holes, dinosaurs, and so on. Newspapers and, in our time, electronic and social media, track the public's interests, and researchers (along with government funding) tend to be responsive to them. In 2018, the most widely followed scientific pursuits were the quest to find exoplanets, especially earthlike planets orbiting other stars, and the successful hunt for the long-sought fundamental subatomic entity, the Higgs boson, along with, as always, new treatments for various cancers.

Plunging into the "consciousness and nature" swamp is part of the third category of popular science, the one best described as "everything else." Even if technogeeks and other informed science lovers have long known that quantum mechanics and other areas of inquiry increasingly point to a bedrock connection between ourselves and the supposedly external and insentient cosmos, wading into this swamp is on very few scientists' to-do lists. The vast majority of scientific inquiries involve hunts for "missing pieces" in clearly defined areas of research. The Higgs

---

† In what may have been the greatest public relations coup of the Renaissance, Nicolaus Copernicus won permanent, unwavering credit for being "the first" to say that Earth goes around the sun and not vice versa, and thus has been forever celebrated as the founder of heliocentrism. In fact, it was someone else—Aristarchus of Samos—who first discovered this (some eighteen hundred years earlier!), even if, strangely, his ultimate reward proved to be anonymity.

was like that, as are searches for alien life and treatments for common medical ailments. In most of science, the questions themselves are easy to frame. And if an answer is found, the accomplishment is readily stated. Consciousness is a more slippery topic—as evidenced by the first question many people ask: *What do you mean by consciousness?* In order to study something, defining it seems a necessary first step, and yet even this is a subject of debate. Thus, most readers will find the subject a major departure from the vox populi issues of mainstream mass-media science.

A study of consciousness requires leaving the world of the known. A study of the connection between consciousness and nature requires us to venture further still into terra incognita. In short, the reader is invited to join us in pushing aside not just the science chaff, but even the sea of compelling, unanswered common questions, to instead dive directly into the center of all experience, the core of all we know, in order to illuminate startling truths about our place in the cosmos.

We'll see that, in innumerable ways, science consistently points to a biocentric interpretation of the universe. As laid out in our first book, *Biocentrism*, we followed this evidence to arrive at a set of seven principles, laid out below, that encompass this biocentric theory of reality.

PRINCIPLES OF BIOCENTRISM

First principle of biocentrism: What we perceive as reality is a process that involves our consciousness. An external reality, if it existed, would by definition have to exist in the framework of space and time. But space and time are not independent realities but rather tools of the human and animal mind.

Regardless of whether you believe there is a "real world out there," a long list of experiments shows that the properties of matter—indeed, the structure of spacetime itself—depend on the observer, and on consciousness in particular.

Second principle of biocentrism: Our external and internal perceptions are inextricably intertwined. They are different sides of the same coin and cannot be divorced from one another.

Aside from the experimental findings of quantum theory, basic biology makes it clear that what appears "out there" is actually a construction—a whirl of neural-electrical activity—occurring in the brain.

Third principle of biocentrism: The behavior of subatomic particles—indeed all particles and objects—is inextricably linked to the presence of an observer. Absent a conscious observer, they at best exist in an undetermined state of probability waves.

This discovery astonished even the physicists who uncovered it a century ago. But experiments have repeatedly shown that how and where particles appear strictly depends on how and whether they're being viewed.

Fourth principle of biocentrism: Without consciousness, "matter" dwells in an undetermined state of probability. Any universe that could have preceded consciousness only existed in a probability state.

Quantum mechanics consistently and accurately predicts how and where the basic particles of matter will appear, with the amazing revelation that prior to observation, they exist in all possible places at once—dwelling in a sort of blurry probability state that physicists call "an uncollapsed wave function."

Fifth principle of biocentrism: The structure of the universe is explainable only through biocentrism because the universe is fine-tuned for life—which makes perfect sense as life creates the universe, not the other way around. The "universe" is simply the complete spatiotemporal logic of the self.

Strong evidence for this is seen in every science textbook table listing the physical constants of the universe. All are perfectly "set" within a fraction of a percent at values that allow complex life-friendly atoms to form, energy-giving stars to shine, and all the myriad conditions that let you now read this to prevail. The laws and conditions of the universe allow for the observer because the observer generates them.

Sixth principle of biocentrism: Time does not have a real existence outside of animal sense perception. It is the process by which we perceive changes in the universe.

Scientists have found no place for time in Newton's laws, Einstein's relativity, or quantum equations. Indeed, even the "before" and "after" reasoning that we call time requires an observer to contemplate some specific event to which others are then compared. As we'll see in later chapters, time does not exist "out there," ticking away from past to future, but rather is an emergent property that depends on the observer's ability to preserve information about experienced events; a "brainless" observer does not experience time.

Seventh principle of biocentrism: Space, like time, is not an object or a thing. Space is another form of our animal understanding and does not have an independent reality. We carry space and time around with us like turtles with shells. Thus, there is no absolute self-existing matrix in which physical events occur independent of life.

Experiments consistently show that distances mutate depending on a multitude of relativist conditions, so that no inviolable distance exists anywhere, between anything and anything else. Indeed, quantum theory casts serious doubt about whether even far-apart bodies are truly and fully separated. Objects cross space in zero time via "tunneling," and can convey instantaneous "information" thanks to the phenomenon of entanglement. Obviously, traversing a million light-years' worth of space in zero time would not be possible if space had any sort of actual physical reality.

As you can see, the principles build upon and reinforce one another. We'll be delving into the science behind them throughout this book, but if they are new to you, at some point it might not be a bad idea to visit our explanations of how each principle is ineluctably derived, as presented in nontechnical language in both the previous biocentrism books. They are quickly reviewed here to allow readers to use them as a springboard for the science that will follow. And, indeed, to serve as preparation for four additional principles that will emerge later in this volume.

But let's not get ahead of ourselves. For now, to understand this whole business properly, we'll move forward by going back. Let's rewind a few centuries and see how the seemingly independent machinations of nature were first found to be linked to us as observers.

NEWTON'S APPLE COMPUTER AND ALTERNATE REALITIES

2

Bodies persist in their Motion or Rest, receive Motion in proportion to the Force impressing it, and resist as much as they are resisted. By this Principle alone there never could have been any Motion in the World. Some other Principle was necessary for putting Bodies into Motion.
— **Sir Isaac Newton**

A t one time or another in our lives, many of us have enjoyed the same fantasy—of magically traveling back in history and meeting a favorite early scientist or visionary. Wouldn't it be fun to hang out with Jules Verne or H. G. Wells and show them photos of modern aircraft and rockets, and tell them that they were right? Wouldn't they marvel at how, in the fullness of time, their greatest fantasies were not just realized, but far surpassed by human technology?

As we probe the workings of the universe, aided by our twenty-first-century computers, we do seem closer to fundamental answers than ever.

Yet we're still in awe of the foundational leaps made by the great minds of the past few centuries. So let's be time travelers and look in on the game-changing breakthroughs that began appearing at a very specific time four centuries ago.

By the Renaissance, increasing numbers of Europeans and many in Asia had grown dissatisfied with attributing all events to the whims of God or gods. They wanted things to make sense rationally. These seventeenth-century rationalists, exemplified by René Descartes, divided the cosmos in various ways, most decisively by separating ourselves as observers from whatever we were contemplating. This subject-object division struck scientists and philosophers of the time as a good and natural idea, since humans were and still are famous for screwing up. Removing the "subjective" aspect of studying nature seemed a prudent first step for avoiding errors.

Also inherent in this new approach to acquiring knowledge was the assumption that past actions are critical to predicting future behavior. This assumption is a useful one when dating, is the logic employed by parole officers, and was key for the physicists of the sixteenth to early twentieth centuries, who relied upon the fact that the trajectory of a moving object was the surest guide to where it would be found in the future.

And it is here, early in the seventeenth century, in an era of challenge and struggle and the devastating visitations of the bubonic plague, that we meet the genius Isaac Newton.

A thin and physically unimpressive man with a hairstyle that would have been right at home during the hippie era of the sixties and seventies, Newton is a pivotal early character in our narrative for two separate, compelling reasons. First, he discovered natural laws that constituted a breakthrough on the most fundamental level by showing that motion obeys the same rules "down here" in our towns and farms and "up there" in the celestial realm—thus tying together Earth and the heavens. Secondly, though it would take centuries to realize it, Newton's laws can also be understood as a peek into alternate realities, a portal to astonishing realizations we will return to later in this book. His insights might have carried him further still had he been able to confront the monster under his bed—the taboo against including the human mind itself in the consideration of how the cosmos operates.

But Newton's laws themselves were no minor upward step in our grasp of the world, and he deserves further accolades for being among the very first to find unity in what had been regarded for millennia as utterly separate dominions—those of celestial bodies and things here on Earth. He put us firmly along the road toward a unified cosmos. Two centuries later, a new generation of brilliant thinkers such as Michael Faraday and James Clerk Maxwell unified other previously seemingly disparate entities—in their case by finding that while magnetism and electricity manifested themselves as distinct phenomena, a single overarching force lay behind them. Yet another half century would bring Albert Einstein, who showed that space and time—as seemingly different as are pizza and laughing gas—were two sides of a single coin. He'd go on to reveal the same *e pluribus unum* motif operating with matter and energy—an unexpected bombshell, for no one had imagined the glow of starlight to be an actual manifestation of material objects converting themselves into energy form. And of course other early-twentieth-century advances in physics and chemistry included the revelation that all elements are composed of identical subatomic particles in a variety of configurations. Increasingly it was looking like a wonderful oneness pervaded nature.

It was Newton who started that ball rolling, and its momentum carries us along with ever-greater speed even today. And by looking more closely at Newton's laws of motion, we can open doors that even Isaac never realized he'd unlocked.

If we begin with his simple examples of a person hurling a stone or an archer shooting an arrow, we realize that what Newton averred is actually quite intuitive. When we were kids and threw snowballs at street signs, we gradually learned how much force to use and how to compensate for gravity's role in the missile's arc, and thus which exact direction to aim in order to hit our target and be rewarded with the metallic ping of success and admiring looks from passersby of the opposite sex.

When we wound up our arms, flexed our biceps, and let the cold sphere fly, a great number of trajectories were available to us:

Fig. 2.1 Different possible trajectories of an object, such as a snowball, thrown from the same position with different velocities, that is, with different speeds in different directions.

This huge range of possible arcs was the result of the force we imparted to the snowball combined with the force of gravity. At the time Newton was developing his laws of motion, this force didn't even have a name—he coined it from the Latin *gravitas*, which meant dignified, serious, or important. By any name, the force that pulled objects toward Earth was always a major player, whether the immediate goal was to win an archery tournament or to accurately hurl cannonballs at a castle we wished to capture. Newton's search for how things moved was motivated by more than a mere desire to make strides as a "Natural Philosopher" (the term "scientist" did not yet exist); it was a very practical quest whose results would improve a great many human endeavors.

As Newton's study of motion led invariably to a probe of gravity itself, he showed that its force is a reliable, invariable quantity that nonetheless alters predictably with changing circumstances: it grows weaker with distance from the earth's center, the force decreasing inverse to the square of that distance—in other words, double the distance between an apple and the earth's center, and the force pulling it toward the ground will be four times weaker. And yes, a falling apple may have started Newton on his gravity quest, or so Isaac himself liked to tell people—though there's no truth to the cartoonish version in which he'd been bonked on the head by one. In any event, it is easy to see how that fundamental fruit, which played such a sinister role in Genesis, might have inspired Newton to formulate his theory. When watching the fall of an apple or anything else that freely plummets, such an object exhibits a predictable trajectory:

Fig. 2.2 Different possible trajectories of an object thrown with the same velocity from different positions.

When gravity's effect is combined with a second force, like the motion of a rock hurled straight ahead from the edge of a cliff, the result is a curving path like the ones in figure 2.2. But for now let's think like Newton and imagine that apple falling straight down from a tree. Nobody is tossing it, so only gravity influences its motion, and thus it heads directly downward at an ever-increasing speed due to the action of gravity. How fast? Well, after one second, the fruit is falling 22 miles per hour—or, if you prefer your apples metric, 9.8 meters per second. If it falls for two seconds, it will be traveling at 44 miles per hour or 19.6 meters per second. After three seconds it would be plummeting 66 miles per hour, fast enough to create applesauce if it splattered onto a rock ledge.

This acceleration is predictable and straightforward. (In practice, air resistance would slow it down a bit, but let's keep everything simple for now.) The closer an object is to the source of gravity, the more strongly gravity acts upon it, and the greater the acceleration of its fall. When Newton said that gravity would grow weaker with distance, he correctly noted that gravity behaves as if all a planet's mass, which he assumed to be the source of its gravity, is concentrated at its very center. Meaning that, gravity-wise, the apple tree on Earth's surface is not at our world's zero point, but is already elevated 4,000 miles up—the distance from the surface to our planet's core.

This was an important bit of fine print, because it enabled Newton to calculate the effect of Earth's gravity on the moon. He knew from trigonometric parallax that the moon's own core was 240,000 miles from Earth's. Meaning, it was roughly 60 times farther from our core than the apple. Therefore, on the moon, Earth's gravity would be 60 × 60 or 3,600 times weaker than the gravity "felt" by the apple. This means that the moon falls at a far slower rate than our terrestrial fruit.

On top of that, the moon isn't just falling straight down apple-like toward Earth. Rather, from the time of its birth, the moon has enjoyed a forward or horizontal motion of 2,290 miles per hour. Thus, just like a hurled snowball, its actual trajectory should be a combination of those two movements—horizontal at 2,290 miles per hour and also gravitationally downward at 0.0060844667144 miles per hour (or 0.00272 m/s^2), which amounts to plummeting 13 feet toward Earth each and every minute.

Here's the fun part. The combination of those two motions yields a lunar path that makes the falling moon sink downward toward Earth at exactly the same rate at which Earth's spherical surface, far below it, curves away and drops off, thanks to the moon's forward movement. As a result, the moon falls completely around the earth, and does so every 27.32166 days. We have a word for when one object is pulled downward toward a heavier body thanks to gravity but also moves sufficiently fast horizontally that it repeatedly circles around it. We say that the object is in *orbit!*

Depending on the object's forward speed, the distance between one celestial body and another, and the strength of gravity (which depends on mass and so varies from body to body), there could be an almost infinite number of possible orbits of one object around another:

Fig. 2.3 Left: A family of different possible trajectories of Earth moving around the sun. Right: When forward motion and gravity are closely balanced, we get another family of possible trajectories—circles. And if we consider various possible distances from the sun to the earth, we see orbits that are concentric to each other. These same rules apply to the moon going around the earth, stars orbiting companion stars, and all manner of celestial combinations observed throughout the cosmos.

Fig. 2.4 Another two possible families of Earth trajectories around the sun: starting from the same position with different directions of velocity (left) and starting from different positions with the same velocity (right).

A major revelation of Newton's was that the moon could have traced many possible trajectories around Earth, just as Earth could have displayed any of an enormous number of possible paths around the sun. The actual trajectories of the moon and Earth are the result of each body's history. A different history would have led to a different orbit; a *very* different history might have led to a drastically different trajectory—for instance, with Earth too close to the sun for life to exist, or the moon so close to Earth that catastrophic tides are a daily occurrence, which would also have made life difficult.

In any case, Newton's laws let us accurately calculate an object's actual trajectory if we know the starting point and velocity (speed and direction)—the so-called initial conditions. These laws are still used by NASA, JPL, and the ESA (European Space Agency) for determining spacecraft trajectories, even if minuscule improvements could be achieved by using the far more complicated field equations of Einstein's relativity. Newton's laws are also used for calculating the future motion of the earth and the moon, which enables accurate predictions of solar and lunar eclipses. And they're used to nail down the future positions of planets, letting us predict phenomena like the transits of Mercury and Venus across the face of the sun.

Yet for all the practical ramifications of Newton's eye-openers, we're most interested in how they set the table, just a bit, for the quantum mechanics that would arrive several lifetimes later. In Newton's day, nobody recognized this potential, since seventeenth-, eighteenth-, and even nineteenth-century physicists had no comprehension of nature's built-in lumpy behavior.

To understand how quantum mechanics finds its roots in the laws Newton developed centuries before—first while dodging the Black Death that was ravishing London and later while relaxing under an apple tree at his farm in the country—we might first backtrack to ponder what paths through space an object would take if absolutely *no* force were tugging on it. If, for example, one hurled a stone in empty space, far from any planet or star.

This is simple. The path would then appear straight, as in figure 2.5:

Fig. 2.5 Possible trajectories in the absence of forces: fixed velocity and variable initial position (left); fixed initial position and variable initial direction (right).

Thus, in cases when no forces are present, an object's motion is very uncomplicated: it moves with a uniform velocity along a straight line. Examples from figure 2.5 let us contemplate two basic families of possible trajectories. One family consists of parallel trajectories, starting from different positions and all having the same velocity. The other family consists of trajectories emerging radially, traveling in different directions from the same central position.

If we add forces back into the picture, we immediately see the resulting influence on the path of the object, because its trajectory will now be curved, and the motion accelerated, thanks to the force acting upon it. This holds for any object in the presence of any force: a planet, spaceship, etc., under the influence of the gravitational force, and, as was found later, also for electrons in the presence of electromagnetic forces.

But let's return to empty space. It turns out that Newton's trajectories, specifically those radiating from a single point, as seen on the right in figure 2.5, behave like the rays in wave fronts.

Meaning . . . ?

Well, to understand a wave front, imagine a pond with calm water into which you toss a pebble. The circular, outward-moving waves that propagate from the impact point determine the so-called *wave fronts*, as shown in figure 2.6. If we draw imaginary straight lines orthogonally (meaning a series of right angles) through these circular wave fronts, we've created "rays," as shown on the right in figure 2.6.

Fig. 2.6 Waves in a pond of calm water (left). Illustration of rays and wave fronts (right).

A century after Newton, the Irish savant mathematician William Rowan Hamilton used this connection between trajectories and wave fronts to create a way to express the motion of a particle as if it were a wave. Newton's laws and their so-called Hamilton-Jacobi reformulation—named for Hamilton's innovations and tweaks introduced by the nineteenth-century math genius Carl Gustav Jacob Jacobi, who was the first Jewish mathematician professor at any German university—let us determine not just how a particle actually moves or will move in the future given current parameters, but also how a particle could have moved had it started from different initial conditions. As we'll see a bit later on, this is at the heart of quantum mechanics, because it is a characteristic of a wave function that it incorporates these alternative possibilities.

It fell to much later thinkers to address the unspoken issue of *why* only one of those possibilities is experienced. A reasoning along such lines inevitably leads to a conclusion that without an observer there cannot be a definite, actually experienced world. After all, it is an observer who determines initial conditions. More precisely, it is an observer's consciousness that is entangled with certain initial conditions rather than

other ones. Thus, initial conditions are intimately connected to the existence of the observer living with just those conditions, as opposed to
some alternative ones, corresponding to an alternate reality.

Whether such alternate universes of "could haves, should haves" can be
considered as actually existing or just as mere possibilities is a matter of vig-
orous debate among experts. But they are a favorite theme in both modern
science and science fiction, and many of us have considered these "what-ifs"
in everyday life, as in this case recounted by author Robert Lanza:

> *I remember attending my thirty-fifth high school reunion with Vicki, one
> of my oldest friends. Memories of her long-dead mother flashed across my
> mind as though they had occurred yesterday. Vicki's mother was a kind,
> self-effacing woman. Her legs were in braces as the result of polio, and
> it was a struggle for her to bring out dessert when I visited. She was the
> mother I always wanted; she always joked that she was going to adopt me.
> Due to her disability, she spent a lot of time watching TV and was always
> watching those fake wrestling matches where they throw people around.
> We chuckled that this frail, gentle woman watched such brutal shows. It
> was Vicki's mom who inspired me to work with Jonas Salk (who developed
> the vaccine that helped eradicate polio from the earth) after college.*
>
> *When I picked Vicki up, I knew her mom would have been thrilled to
> know that we were going to our thirty-fifth high school reunion together.
> If she had still been alive, she would probably have been watching wres-
> tling, and told us some funny story to make us laugh before sending us
> on our way. How proud she would have been of the lawyer and doctor
> Vicki and I had become. It's sad she didn't live to see that future. But I
> like to think that, in some other universe, she did—that as we left for our
> reunion that night, somewhere Vicki's mom leaned back on the sofa and
> watched the rest of the wrestling match with a smile on her face.*

We'll return to the topic of alternate realities at length in Chapter
4. When we do, remember that this idea, though it seems not just thoroughly modern but ripe for the most mind-bending of sci-fi scenarios, in
fact got its start in the days of plague and powdered wigs, with Newton
and his apple.

QUANTUM THEORY CHANGES EVERYTHING

Do not keep saying to yourself, if you can possibly avoid it, "But how can it be like that?" because you will get "down the drain," into a blind alley from which nobody has escaped. Nobody knows how it can be like that.
—**Richard Feynman on quantum mechanics**

We cannot arrive at the revelations of biocentrism without first visiting quantum mechanics. We do this even though it's a land mine. A real can of worms.

On the one hand, quantum theory was such an amazing breakthrough in our understanding of the cosmos that, even now, a century later, physicists refer to all previous science as "classical physics," showing that they felt compelled to establish a major before-and-after distinction, akin to how the widespread adoption of Christianity led much of the world to divide time into separate BC and AD components after the life of Christ. Quantum theory—or "QT" as we'll abbreviate it from here on out—not only paved the way for biocentrism, it created an entirely new way of looking at the world, rewrote the rules governing it, and so transformed science that virtually every technological advance since owes some debt to its insights.

But the "can of worms" aspects are multiple. First, QT often defies logic. So much so that Niels Bohr, one of its founders, said, "Those who are not shocked when they first come across quantum theory cannot possibly have understood it." Half a century later, the famous theoretician Richard Feynman went further: "It is safe to say that *nobody* understands quantum mechanics."

This wasn't because the equations were difficult or the math laborious—it was the concepts themselves. Feynman was merely expressing that to wade even tentatively into QT required that basic assumptions about reality be abandoned. Here's an example:

If we shoot a photon toward a sensor, its arrival will be easily detected. However, we might first bounce it off a kind of beam splitter or two-way mirror so that it can arrive at the detector by taking either of two paths. Call them route A and route B. Well, other detectors along the way show that before smashing into the final sensing device, the photon has taken *neither* path A nor path B. It also has not somehow divided itself and taken both paths, nor has it arrived by taking neither path. Somehow it has avoided the entire setup.

Those are the only choices our logic can entertain. If the rational world is to be trusted, the photon must have experienced one of those four possibilities, as there are no others. Yet, amazingly, the photon nonetheless did *something else*—something other than A, B, both, or neither.

Physicists are now used to this. They even have a name for such illogical behavior, of objects conducting themselves outside of the choices imposed by common sense. They say the photon was in a state of *superposition*—that is, it was free to exercise all four possibilities at once, even if they appear mutually exclusive to us.

Aside from such seeming impossibilities, there's also the fact that our observation—or even knowledge in our minds—changes how physical objects behave. This was the first solid hint that the observer might play more of a role than being a mere witness to nature's pageant.

And QT's "can of worms" goes deeper still. Since quantum phenomena appear to happen instantaneously, and they do not require any travel
time (even that at light speed) to spread from one place to another, the
search for an explanation inevitably raises the idea of universal connectedness. The no-time and no-distance implications resemble the mystical
teachings of Hinduism and Buddhism. This has spurred many writers to
claim that science and religion have now fused and are in agreement about
the fundamentals of the cosmos. While such philosophical or metaphys-
ical musings are not out of place or even inherently wrong per se, a great
number of TV documentaries, books, films, and articles exhibit a serious
misunderstanding of QT. In short, they often get it wrong.

As just one example, QT's unity aspect was the central story line
in the 2004 movie *What the Bleep Do We Know!?*, which took in $10.6
million at the box office. It featured interviews with experts on quantum theory, a few of whom nonetheless made silly claims that bore no
resemblance to the real thing. One asserted, for example, that QT says
that each person can determine her own future in every way. Actually,
the opposite is true: all "predictions" of future events by QT are probabilistic and hence strictly statistical. No one can consciously control
external physical events that affect them—meaning those that inherently
lie beyond human volition such as a boulder rolling down a hill into the
path of their car—any more than they control whether the next coin flip
will come up tails or heads.

Given the importance of quantum theory to our story, and the fact
that the struggle to understand its oddities has led to so much popular
nonsense on the topic, it's probably worth investing a few pages to show
how it began, how it evolved, why it was successful in clarifying areas of
nature that had previously been bewildering, and how it propelled us
toward the novel findings that we'll encounter later in the book.

It all started with light—meaning, the emissions from hot objects.* If
a hot object is examined spectroscopically, we can unscramble the different

---

* The studies actually involved the properties of "black-body radiation." In real life,
an object being illuminated by something else, such as a planet's surface bathed
in sunlight, will reflect away some of the incoming energy, with the reflection percentage expressed as the object's *albedo*, determined by the darkness of its surface,

wavelengths of energy that come from it, which may include visible colors making up the glow we associate with, for instance, a red-hot iron poker, as well as invisible emissions such as infrared. According to the properties of the waves making up various frequencies of light and the laws of classical physics governing the way heat energy is divided up, any hot object should emit a certain quantity of weak red and infrared light, a greater amount of more energetic green light, and an almost infinite quantity of small-wavelength, high-energy light in the violet and especially ultraviolet range.

But that's not what happens. Instead, a peak quantity of light is emitted at a specific wavelength, an exact color that depends solely on the object's temperature. Classical physics couldn't explain what we see.

In 1900, German physicist Max Planck found a way to make the math match the experimental results by proposing that the atoms making up a glowing object absorb and emit light of various frequencies only in multiples of some fundamental unit. Here was the introduction of the idea of a "quantum" (or specific amount—"quantum" comes from the Latin for "how much") of energy, and the first big milestone of quantum theory.

In 1913, Niels Bohr used the "discrete quanta" idea to explain why atoms continue to exist, when classical physics insisted that they should all self-destruct. Specifically, the Danish physicist pointed out that as electrons rush around in their circular orbits, classical laws say they should be emitting electromagnetic waves every trillionth of a second. The accumulated energy loss should soon make them spiral down into the proton at the atom's center. But—thankfully for our continued existence as human bodies on a stable planet—this doesn't happen.

---

whether it's smooth or ripply, and other factors. Some objects, like hailstones, might also transmit some of the energy through their body and out the other side. But a black body is a theoretical object that reflects nothing and absorbs everything, no matter what angle the energy comes from or what frequency the energy's waves. Thus, depending solely on its temperature, it will radiate away the absorbed energy in a very specific way. The issue in the late nineteenth century into the early twentieth was that classical science's predictions about the nature of this emitted black-body radiation were shown to be wildly incorrect, strongly indicating that physics needed a major upgrade.

You may recall from high school that light is created when an electron orbiting the nucleus of an atom jumps or falls inward, releasing a bit of energy as it moves to a shorter orbit. A common way that many attempt to picture this is by visualizing planets orbiting the sun. If Earth were somehow suddenly gifted with extra energy, it might use this energy to overcome some of the sun's gravity and jump outward to a larger orbit. Depending upon the amount of "extra" energy, this new orbit might be just a few miles farther from the sun than we are presently positioned. Or it could be a million miles farther. Or ten million. Or anywhere in between.

The same was assumed to be true of electrons. But building on Planck's quantum innovation, Bohr suggested that each electron must remain in a discrete orbit with a fixed radius from its nucleus. He proposed that an electron was only "allowed" to be located at a particular distance from its atom's nucleus, or at another particular distance, but nowhere in between.[†]

If an electron was hit with a bit of energy, it would jump to a larger orbit, *but it would have to be a specific jump.* And to do so, it could absorb a particular amount (or quantum) of energy, but not less and not more. That amount of energy would then be subtracted from the energy source, which would leave a telltale black vacancy, a gap in its spectrum.

Having gained energy, that electron could then give it back by falling inward to the next lower orbital state while simultaneously emitting a precise amount of energy—and thus a precise color of light. The discrete bits (quanta) of energy were specified as h, called the Planck constant. And all "jumps" of energy had to be multiples of this number.

The Planck unit of energy is no arbitrary thing; it is a constant that is observed in nature throughout the cosmos. Planck himself accurately determined its value through observation and experimentation. It became a brand-new fundamental unit in physics.[‡]

---

† One result of this discovery is that we could finally know the size of the atom. It's 0.0529 nanometer, or about 1/200th of an angstrom in width.

‡ The value of this constant is $h = 6.6218 \times 10-34\text{J-s}$ (where J represents the joule, a standard unit of energy). In practice, a number that keeps arising in nature is h divided by pi doubled, so *this* value, expressed as "h-bar," is what is most often

But it was also very strange from the get-go. Imagine if celestial bodies behaved this way. Imagine if the moon could orbit Earth where it is now, or at twice its distance, or at three times that distance, but in no in-between position—not because of other planets or objects acting upon it but . . . just because. Now imagine if it jumped from one of these orbits to another in zero time. And if, when doing so, *it never passed through the intervening space.* Yet this is exactly how electrons behave, making discrete jumps that somehow avoid any transitioning in space, and occurring while no time elapses.

So, yes, the spectra of hot objects and the continued existence of atoms now made sense. But this came at a cost: the new understanding defied rationality and previous experience, and even Planck himself struggled with it. Years later, he admitted, "A new scientific truth does not triumph by convincing its opponents and making them see the light, but rather because its opponents eventually die, and a new generation grows up that is familiar with it."

Planck's introduction of the quantum in 1900 changed everything, and it was only the beginning. Just five years later, in 1905, Einstein applied the quantum principle to the substance of light itself. In short, he said that light, long known to be a wave, was also composed of clumps or discrete energy packets—essentially particles of light, called *photons.* This particle nature was fully confirmed in 1922 when it was shown that light scattering—the phenomenon that gives us our blue sky—could only be caused by light acting not as waves, but in its particle form.

Then, in 1924, French physicist Louis De Broglie used the existing quantum laws to show that it was not only light that had a wave and also a particle guise. Every particle in the universe is also a wave and enjoys the same dual nature. De Broglie built on the work of Planck and Einstein to devise a formula to describe the wavelength and energy value for objects of various sizes. De Broglie's conclusion that all particles like

---

used by physicists, who often multiply h-bar by the angular frequency of a particular color of light. "H-bar times angular frequency" is equal to the "packet" or discrete bundle of energy that Einstein would soon call a photon—the particle aspect of light! Everything was starting to tie together!

electrons also have a wave nature was proven by actual experimentation just two years later, using diffraction effects on crystals.

Unfortunately (or fortunately, for those who enjoy strange and unexpected discoveries), one bizarre revelation seemed invariably to lead to another, as if science was now wandering through a series of Wonderland looking glasses. Though each problem that presented itself for investigation was logical, the answers were anything but. Thus, in the 1920s, physicists found themselves stunned and exhilarated as they passed through these strange new doorways, each leading to new breakthroughs in our understanding of nature. Along the way, they had to tackle afresh such seemingly straightforward subjects as how to determine where any given particle or bit of light is located. This sounds like a simple enough inquiry. For, inarguably, if anything that is a wave (meaning, everything in the cosmos) must also possess a particle nature, then surely, like all particles, it must also have a position at any given moment. It must be somewhere, and nowhere else. But how to determine whether it's here or there? Scientists reasoned that if an atom is a cluster of waves, then by looking at the way these waves interfere with one another, it should be possible to identify harmonic beats, places where individual waves are not canceling each other out, but instead acting in reinforcement. This yields a statistical "spread" of these places that tells us where any given particle is likeliest to be. All such predictions soon proved to be spot-on. But "where it is likeliest to be" was as close as they could get.

Then in 1927, Werner Heisenberg introduced his now-famous Uncertainty Principle, which explains mathematically why any object with a wave nature (meaning everything, but most specifically tiny objects) has a built-in limitation on what we can know about where it is located and how it is moving. It isn't just that we observers contaminate or influence what is being seen (which is how many initially regarded the uncertainty business for decades afterward), or that any interaction between classical and quantum-sized objects causes such uncertainty; rather, it's an inherent attribute of waveforms. This uncertainty applies to all pairs of properties that are linked with each other in a certain way. The bottom line is that the more precisely we know how an object is moving, the less accurately we can know where it is located at any given moment.[§]

This has far-reaching consequences. Remember how Niels Bohr used a quantum model of the atom to explain why electrons didn't go crashing into protons, as classical physics said they should? Well, Heisenberg's Uncertainty Principle offered another explanation. If the electron crashed into the nucleus, we'd know that its motion was now zero. And we'd also know its location: it's right there in the center of the atom! But since Heisenberg's principle insists that we cannot know both position and momentum with precision, this event simply cannot happen. And it doesn't!

We've seen how, during the first three decades of the twentieth century, farsighted physicists Max Planck, Albert Einstein, Louis de Broglie, Niels Bohr, and Werner Heisenberg—followed soon by the likes of Erwin Schrödinger and Paul Dirac—created mathematical models with unprecedented predictive power that explained nature's bewildering oddities and showed us how things operate on the smallest scales, those that comprise the cellular level of the universe. And they all soon won Nobel prizes for their troubles. They used statistical methods and discovered astounding "constants" that showed us nature operates differently on the submicroscopic level than it does in the macrocosmic world visible to our eyes. We call the entire body of their work quantum theory or *quantum mechanics*. It may be labeled a "theory," but QT has passed every observational test thrown its way.

It also made several specific predictions that seemed outright impossible, and it is these that most link quantum theory with biocentrism in general and our latest refinements of it in particular. One of these predictions has to do with what has come to be known as *entanglement*.

In 1935, Einstein and two other physicists, Nathan Rosen and Boris Podolsky, addressed a curious quantum prediction about particles or bits of light that are created together, which are said to be "entangled." We can shoot a photon or bit of light into a crystal of beta barium borate, for

---

§ An easy-to-grasp biocentric argument for this uncertainty can be found in each of the previous biocentrism books.

Fig. 3.1 Our understandable desire to picture atoms and their encircling electrons unfortunately comes up short. We use the term "orbital" to describe an electron's location, and this might suggest it's circling around the atom's nucleus like a planet orbiting the sun. But an electron does not actually orbit. Instead, it might be pictured as having some likeliest distance from the nucleus, somewhere in a spherical shell. But it cannot have its position in that shell pinned down at any given moment. If we plot the probabilities of where we might find this electron, the black areas show the highest likelihoods, while white areas show where it's not likely to be found.

example, and find that two photons emerge. Each will have a wavelength twice as long as that of the one photon that went in, meaning each has half the energy, so that all together the energy coming out is the same as the energy going in, as dictated by the laws of physics—quantum or otherwise. What's bizarre is that, according to QT, each of these now-entangled photons, even if they fly off at light speed to become widely separated, must always somehow "know" what the other is doing, and "respond" with complementary actions of its own. For example, if one photon's waves are observed to be vibrating in a horizontal direction, its twin would know about that observation and exhibit a complementary property—in this case a vertical polarization. Indeed, quantum theory said that this "knowledge" would be instantaneous even if the pair were separated by light-years. Which in turn would mean that the seemingly ironclad rule that Einstein himself had discovered, of light speed as the fastest speed in the universe, would fly out the window.

This was too much to accept, which is why Einstein, Podolsky, and Rosen now argued that such simultaneous behavior would have to be caused by unknown local effects—such as a force we weren't yet aware of or contamination of the experiment—rather than, as they pejoratively put it, some sort of "spooky action at a distance."

There was a second troubling issue highlighted by this prediction. Why should an observation of the first photon be at the root of any behavior in the first place? What difference should it make if someone takes a look at that bit of light? Doesn't it possess its properties (of polarization, say) independent of its being observed? The amazed physicists of the early twentieth century were discovering the answer was, "Not really."

Essentially, QT tells us that until observed, particles and bits of light exist only as a kind of energy blob of blurry possibility, a mathematical probability having such-and-such likelihood of being this, and such-and-such likelihood of being that. Upon observation, a group of particles or bits of light will indeed each materialize in accordance with their mathematical probabilities, shedding their fuzzy wave nature to manifest as discrete objects that behave like particles or waves, depending on the experiment used to detect them. Einstein hated this prediction because it suggested that reality wasn't definite, it was probabilistic—like a game of chance, which is why these quantum predictions inspired his famous sneering comment of "God does not play dice!"

Declaring that "you can't get something from nothing," many to this day naturally ask what that pre-manifestation "blob of possibility" actually is: What was there before that photon or electron snapped into definite existence? The term used then and now is to speak of its preexistence as being or having a "wave function." (As we'll see, this is a bit iffy from the get-go because much evidence suggests the photon or particle simply didn't exist prior to observation, so that we're essentially trying to find a label for what amounts to nonexistence.) When an object materializes, it does so according to the probabilities described by this wave function; we might think of wave function as simply a mathematical likelihood. But is a "likelihood" a real thing, or merely a human concept we use to describe it? We'll talk at length about wave functions in the next chapter, but any perusal of modern quantum physics texts reveals that the label continues to remain vague and mysterious, with physicists themselves unsure of what a wave function really is: Some actual energy object? Some sort of probabilistic ghostlike entity? One thing seems certain: Upon observation, an object's wave function "collapses" (to retain the term that's been favored for more than a half century), which is simply a way of saying that the object then becomes a specific entity with real physical characteristics, and—starting at that moment—will then continue its existence indefinitely.

The "collapse of the wave function" is therefore a material object's birth moment.

At that point, if it's an electron, it may be observed to have a vertical spin. If it's a photon, it may be seen to have a horizontal polarization, meaning the electric component of its waves is oscillating side to side rather than up and down. The point is, upon observation, this object exhibits definite physical characteristics that are not transient but which endure until disturbed by some other interaction.

But back to entanglement. The prediction about the behavior of entangled particles rests on the fact that twin particles created from a single particle in this way would share a wave function. The two photons may fly off at the speed of light and have independent lives, perhaps for millions of years. But if one is observed and found, say, to have a vertical polarization, the distant photon (or wave function glob or however we want to picture it) instantaneously "knows" that its twin was observed, and it, too, collapses into a photon—with perfectly complementary properties, in this case a horizontal polarization. *Together they constitute a matching set.*

"Impossible!" said Einstein, Podolsky, and Rosen. To them, this prediction was proof of a flaw in quantum theory. They went on to address entanglement with such depth (and scorn) that the phenomenon was known forever afterward as an "EPR Correlation," using the first letter from each of the physicists' names.

But experiments performed since, attempting to clarify the bewildering predictions about entanglement, show that Einstein was wrong. Specifically, following somewhat inconclusive (though intriguingly suggestive) experiments performed in 1972 by Stuart Freedman and John Clauser, and in the early 1980s by Vittorio Rapisarda and Alain Aspect, a Geneva researcher named Nicolas Gisin managed a decisive demonstration of this behavior in 1997. He created pairs of entangled photons and sent them flying apart along optical fibers. When one encountered the researcher's mirrors and was forced to make a random choice as to whether to go one way or the other, its entangled twin, seven miles away, always made the complementary choice instantaneously.

Experimental evidence that a photon can "decide" how to be or act based on the actions of another photon separated by a great distance is of course fascinating. But certainly one of the most remarkable aspects of the experiment was the word "instantaneously."

Remember, one of Einstein and company's chief arguments against the possibility of this behavior was that nothing can exceed the speed of light. Even when black holes collide to create awesome gravitational ripples spreading across the cosmos, the effect is strictly confined to this inviolable speed limit of 186,282 miles per second. Yet that speed limit didn't seem to apply in Gisin's laboratory. The reaction of the entangled twins in 1997 was not delayed by the amount of time light would have taken to traverse the seven miles between them—it happened at least ten thousand times faster, which was the equipment's testing limit. The echoed behavior was presumably simultaneous.

The mounting experimental evidence on entanglement was so bizarre, it drove other physicists to a frantic search for loopholes; some insisted previous experiments might have introduced a bias, for instance by being more likely to detect linked particle events. These criticisms were silenced in 2001, when, as reported in the journal *Nature*, National

Institute of Standards and Technology researcher David Wineland used beryllium ions and a setup with a very high detector efficiency to observe a large enough percentage of in-sync events to seal the case.

So this fantastic behavior is real. But how can it be? That year, Wineland, who won the Nobel Prize in physics a decade later, told one of the authors, "Well, I guess there really *is* some sort of spooky action at a distance." Of course, as he knew, this explains nothing.

To sum up, particles and photons—matter and energy—go from blurry, probabilistic, not-quite-real "wave function" statistical entities to actual objects the moment we observe them. And they can transmit knowledge of their newly acquired state clear across the cosmos, causing an entangled "twin" to assume complementary attributes instantly, in real time. Or maybe that's not what happens. Perhaps no entity "sends" information, nor does any other one receive it. Perhaps, instead, both simultaneously spring into existence when either is observed. Whatever the specifics, our logic struggles to play catch-up. Among the implications:

a.   Neither space nor time actually exist. Because surely if space has any kind of reality, then traversing it would take time, even if just a little bit.

b.   There's some kind of unity to the cosmos, a connectedness outside of space and time.

c.   The act of observation is somehow central to the existence of reality.

Spooky or not, entanglement undoubtedly exists in the quantum realm. But whether or not the laws of quantum mechanics "scale-up" to the macroscopic objects surrounding us—and how to detect it if they do—is another matter, one that has been pondered by researchers for decades. In 2011, an international team of scientists from Oxford University, the National University of Singapore, and the National Research Council of Canada conceived an experiment to see if the quantum concept of entanglement extended to the everyday realm. They focused on a pair of three-millimeter-wide diamond crystals—about the size of the diamonds in a nice pair of earrings—objects that were not even microscopic, much less subatomic.

The scientists induced vibrations in one of the diamonds, creating a phonon—a unit of vibrational energy. Because of the design of the experiment, there was no way of knowing whether the phonon had been made to vibrate in the left-hand diamond or the right-hand diamond. The researchers used laser pulses to detect the phonon, and the pulses showed that the phonon came from both diamonds, rather than one or the other. The diamonds were entangled! Apparently they were sharing one phonon between the two of them, even though they were separated by a distance of about fifteen centimeters.

In 2018, an article in *Scientific American* resurrected the question, saying, "Scientists have wondered where exactly the microscopic and macroscopic worlds cross over . . . the big question being whether quantum effects play a role in how living things work." The article was discussing a 2017 finding published in the *Journal of Physics Communications*, the result of research by a group at the University of Oxford.

By observing photosynthesis within microbes, the Oxford group claimed the first successful entanglement of bacteria with photons—particles of light. Led by quantum physicist Chiara Marletto, they analyzed a 2016 experiment performed by David Coles and his colleagues from the University of Sheffield, in which Coles isolated and sequestered several hundred photosynthetic bacteria between two mirrors. By bouncing light between the mirrors, the researchers caused a coupling or connection between the photosynthetic molecules within six of the bacteria. In this case, the bacteria continuously absorbed, emitted, and reabsorbed the bouncing photons, exhibiting the kind of simultaneous behavior unseen in classical science.

In short, today's science has carried the bizarre activity of the quantum realm, first uncovered a century ago, into the macroscopic and biological world. *Our* world!

Now you understand why we had to visit quantum theory. It not only constituted vast advances in human knowledge, it also laid down stepping-stones that later theorists would use to travel even further—from the quantum world to our own, and from our own, as we'll see, to the possibility of many others.

INTIMATIONS OF IMMORTALITY

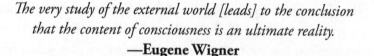

4

The very study of the external world [leads] to the conclusion that the content of consciousness is an ultimate reality.
—**Eugene Wigner**

From Newton through the rise of quantum theory, we've explored the roots of biocentrism's central premise—that we, as observers, create reality. Now it's time to dive into what that really means and how it happens. And, to do that, we need to take a closer look at a key concept from the previous chapter, that moment in which the possible becomes the actual: the collapse of the wave function.

We saw that quantum mechanics describes the motion of a particle in terms of a *wave function*—a term that expresses the blurry, not-yet-definite preexistence of all quantum entities, whether particles of matter or photons of light. Since this term is important, even if it has now bewildered four generations of laypeople attempting to fully grasp it, we'll start by splitting it into its halves and making sure we understand what is meant by both "wave" and "function."

In its simplest form, a wave is a disturbance in some matrix like air or water, through which energy travels from one place to another. Types of waves might be distinguished by how they travel—whether they wave up and down like an ocean wave, or side to side like a rope snapped horizontally. Or, alternatively, waves can be grouped according to what mediums they can travel through. Longitudinal (vertical) waves can pass through liquids and gasses while transverse (sideways) waves require the material to be solid.

We'll keep talking about waves a bit longer, but meanwhile, let's sneak in that second half of the term "wave function" by defining "function." This one is easy. A function is a mathematical way of expressing a relationship. Consider, for instance, a familiar graph of temperature versus time, which is straightforward enough: The afternoon temperature is likely to be higher than it was in the morning. But temperature also depends on the location—it varies from place to place, and is thus a function of position. The same holds for the height of a wavy water surface, which also varies from place to place. Mathematicians could use the formula "$y = \sin x$" to describe the frozen-in-this-moment appearance of a wave produced by throwing a rock into a pond. But since this shape is constantly in motion along the water surface, which we could "graph" or visualize in our minds as the x-axis, we'd want to bring time into the picture by using "$y = \sin(x - t)$." Don't worry about the equation here—the point is simply this: a wave function is a mathematical representation of a wave, one that can describe motion. That is, it doesn't just tell us what a wave's shape looks like now, but it also includes the way it changes over time.

All of this interests us because the universe is composed of countless particles that, as we've seen, have "wave nature." Specifically, 10 followed by eighty-four zeros—that's the number of subatomic particles like electrons in the universe. Then there are the photons—bits of light, which we might think of as morsels of energy. There are roughly a billion times more photons in the cosmos than there are "solid" subatomic particles like electrons. And all these multitudinous point-like objects, whether electrons or photons, travel in ways that a wave function can describe! So if we want to know what's going on—where anything is or how it moves—we need to stick with waves.

You may recall from Chapter 2 that an object like an electron can be represented as moving like a straight-line ray orthogonally to (at right angles to) the spreading, curving "front" of a traveling wave.

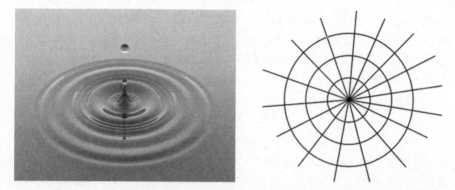

Fig. 4.1 Waves created by a water drop (left). Illustration of rays and wave fronts (right). In the example of an object dropped into water, the circular, outward-moving waves that propagate from the impact point determine the so-called "wave fronts."

The rather complicated expression describing this shape of movement is the wave function, and in quantum mechanics, "wave function" has its own symbol, the Greek lowercase letter psi: ψ. The wave function of a quantum particle describes a wave like the one we see rippling in water in the figure above, and the rays moving orthogonally to its wave fronts are possible trajectories of the particle.

The wave function of an object like an electron describes the probability of observing it at a certain position, and that's actually everything we can possibly know about the object. In practice, unlike with macroscopic objects we can see, which have actual defined trajectories, the future motion of the myriad tiny particles that make up the universe can only be given as a probability. So for all our trouble, the wave function equation can't reveal exactly where an electron is or how it is moving. Instead it gives us the probabilities for such things, which we've learned to accept as "good enough."

A wave function thus carries information, however fuzzy, about possible positions of a particle. But not all those possible positions will actually be experienced by us. Unobserved, a particle's wave function may spread over a vast realm of possible locations, but after we've made an observation, the wave function loses that wide range of freedom and automatically becomes closely concentrated around a specific position; after all, we just saw it. This transition from a wide to a narrow wave function is called *wave function collapse*. And this true eureka moment in the life of a particle or bit of light—its birth moment—is when it leaves behind its strange and illogical attributes to assume the guise of a single well-behaved object that has no more mystery than a cheeseburger.

Remember, in Quantumland, the realm of the tiny, a particle like an electron exists in a state called *superposition*. Meaning, it is doing everything that is possible at once. It is on highway A, on highway B, on both, and on neither, all at the same time. We think of an electron in superposition as having multiple contradictory states existing simultaneously, such as both an up spin and a down spin. In reality, the orientation of spin, for example, is always mutually exclusive; an electron cannot display both—and, indeed, is always found to have one or the other—when it is measured. *Before* it is measured, however, you can't speak of an electron as having any definite properties at all.

The macroscopic world doesn't act that way. Your room's light is either on or off, but it is not both, and certainly not neither. We might watch a powerfully hit baseball as it zooms toward the outfield. Its path is clear. It is not taking two paths at once, one foul and one fair. It is either fair or foul, but not both. It either flies high as a pop-up, or whizzes fast and low, but not both at the same time. That wouldn't even make sense (and it would be very confusing for the umpires)!

So, for a full century, physicists have wondered what causes an object's behavior to make the switch from the anything-goes quantum realm to the commonsense classical science realm when it is measured. What is it, exactly, that causes the wave function to collapse so that the object obtains a real-life attribute? If it was in an anything-goes state but then becomes a real item upon observation, it seems only logical that the observation is what caused the wave function to collapse . . . but if so, how? And on the other hand, correspondence is not causality. There is a 100 percent correlation of day always beginning just as night fades away. Yet, despite this invariable link, night is not the cause of day. But if observation doesn't cause the wave function to collapse, what does?

Throwing our rational minds for a loop are all the tricky aspects that arise when dealing with tiny objects—phenomena we needn't think about when it comes to calculating the position of, say, the moon. One of them is that measuring or even observing a subatomic object always has an effect on it because any information we gain always involves an energy exchange. Think about it: If you see something, it means photons or bits of electromagnetic energy have impacted retinal cells, delivering the electromagnetic force, one of the four fundamental forces, to atoms in those cells, and ultimately caused electrical impulses to arise. What can you ever perceive without an energy exchange? The mere process of observation can alter what's occurring at a fundamental level without you even being aware of it—just as using a flashlight to try to learn what mice do at night will change their nocturnal behavior and automatically result in erroneous conclusions.

Therefore, the question of exactly how and why an observer "causes" things to be the way they are may be both the issue that most demands our attention and the most stubbornly difficult to untangle.

Countless experiments have yielded clues, among them the famous double-slit experiment in which electrons are beamed toward two close-together openings in a barrier. If the beam is sufficiently wide so that the electron has a fifty-fifty chance of passing through either hole, we've set up an interesting situation. We know that, according to the rules of the quantum world, each electron in the beam exists as a blurry wave function. Therefore it will experience all possibilities at once and pass through both openings. Then those different portions of the electron wave "interfere" with each other and create a distinct, easily discernable interference pattern on the detector screen at the back of the experiment.

But now, step into the lab and repeat the experiment, this time adding in a measuring device that tells you which slit the electron passes through. And just like that and all by itself, the electron loses its blurry existence as a wide probability wave passing through both holes and behaves instead like a particle, passing through one hole only. Now there's no interference pattern on the screen.

Variations of this experiment have been performed a gazillion times in the past seventy-five years. The single variable that always accomplishes the collapse of the wave function—or the electron's transition from fuzzy wave behavior to classical particle behavior—is the observation, or the measurement by an observer. In some variations, the only thing that changed from one version of the experiment to the other was *information in the observer's mind*! In that case, when the final detector was set up so that a computer scrambled and randomized the results to make them unintelligible, the electron then retained its quantum behavior and passed through both slits, producing the interference pattern. But whenever the scrambler is turned off so the observer gets valid information about which slit or slits are penetrated, then, in that nanosecond, the interference pattern vanishes and the electron's path reverts to a single slit—even retroactively! It's a particle or a wave, with its path clearly changing from "both slits" to "one slit" depending solely on what the person in the room knows! It's quite spooky.

There is no avoiding it. Somehow, observation is the cause of the quantum/classical transition. Now, all sorts of other explanations have been tried. They include the idea that a particle acting in a quantum way can lose its anything-goes, all-things-possible, quantum characteristics due to wave interference when merely put in the company of macroscopic objects and suffering their influence. Others have ventured that perhaps it's a gravitational field that does it. But in all cases, some problem has been found. Even today, debates continue about whether the observer must be a living, conscious being. Many aver that any interaction or measurement "forces" a photon or subatomic particle to assume definite properties and thus counts as an observation, collapsing its wave function. In truth, some properties of observers are enough to invoke some physical effects, while other properties lead to other effects. It is hard to untangle this for many reasons, including some that probably seem obvious: Sure, we can make measurements via automated instruments. But all observations (even measurements made by instruments) can only be known to us by consciousness. If nobody ever looks at the results, the whole issue stays blurry and speculative. What's more, as we will see in Chapter 11, it turns out that an observer with memory is required in order to establish an arrow of time—and correspondingly, the relationships of cause and effect in everything we observe around us. (Appendix 1 contains further discussion of the observer issue, for those who want it.)

In the final analysis, we can only say for *sure* that a conscious observer does indeed collapse a quantum wave function. And needless to say, the implications of this go deeper than almost anyone first imagined—as we're about to see.

A wave function will typically spread over a large range of possible locations, but after we've made an observation, the measurement loses this wide range of freedom and automatically becomes closely concentrated around a specific position—as mentioned, this transition from a wide to a narrow wave function is known as wave function collapse.

Let's watch wave function collapse in action. Picture the wave function of a single particle, say an electron, propagating as a plane wave as illustrated in figure 4.2. If it helps, think back to our "ripple in a pond" wave. The planes are like the ripples of the traveling wave fronts. The rays, which are not shown in the figure below, are orthogonal lines to

Fig. 4.2 A plane wave function interacts with a fluorescent screen. When an observer looks at the screen, he sees the spot, which can be anywhere on the screen.

those planes. (The waving line in the figure only serves to remind us that a wave is heading toward the screen.)

If we put a fluorescent screen in the electron's way, then after looking at the screen, we'll observe a single spot somewhere on it.

The probability of observing the electron (or whatever particle) at a given position is determined by its wave function. (In practice, physicists derive the mathematical probability by taking the wave function's square.* Remember, we are merely describing the process and sparing you, dear reader, from actually following the math.) Before we observed it, the probability of the electron arriving at a certain point on the screen was the same for all points—if another electron's wave arrives at the screen, we'll then see a different spot, most likely at some other place on the screen, and after many such encounters, we'd observe a uniform distribution of spots on the screen.

Before we look at the screen, when we still have zero information about the particle's position, the wave function is spread over all space as a plane wave. But once we look and see a spot, we have useful, finite information about the question *"Where is the particle?"* and the wave function collapses to become localized like a cloud around a certain position, as in figure 4.3.

Thus, one easy way to understand this whole business is to regard wave function as a delivery method for information about likelihoods and probabilities. It tells us where a particle will most likely materialize—and, conversely, where we needn't bother looking for it. When the wave function is no longer vaguely spread all over the place in a "(virtually) anything's possible" plane-wave situation, but instead is usefully localized, as shown earlier, we know we are homing in on an answer to our *"Where is this thing?"* question.

So far, we've been considering the wave function of a single particle. But when describing a system of two, three, or many particles, not to mention the entire universe, the wave function is an expression of the

---

* More precisely, wave function is a two component object (ψ_1, ψ_2), which can be written as a complex number $\psi = \psi_1 + i\psi_2$, whose absolute square, $|\psi|^2 = \psi_1^2 + \psi_2^2$, gives the probability.

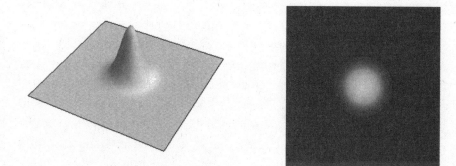

Fig. 4.3 The location of the probability density calculated from a wave function localized around a point. Such wave function gives us the information that the particle will be most likely found in the center of the "cloud."

positions of all those particles. If we have sufficient computer power to handle the math, such a wave function will then give us the information about what the universe that we experience looks like, and what will most likely happen in the next moment.

Wave function thus represents the world experienced by an observer such as you. But the world contains other observers, too.

We've seen that a wave function describes probability. But when we ponder the real world containing numerous observers, we are forced to expand our understanding of "probability." After all, is a probability the same for every one of us? Not necessarily. Every card player knows that the probability of whether another player has a certain card changes according to the information gained during play. And since players have different cards in their hands, that probability calculation is different for each of them. So figuring out how situations unfold or particles arise or movements interact becomes much more complex when we consider the real-life multiplicity of the actual reality around us, and the simple term "wave function" suddenly demands involved computations and intense computer resources. (Of course, the reader will be spared all such mathematical hardship.)

This world of many observers brings us at last to a discussion of the "theory of many worlds." In an experiment such as that shown in figure

4.2, before you looked at the screen, all positions of the spot were possible, and the wave function of the screen was a superposition of all those possibilities. When you look at the screen and see the black spot marking the electron's impact, that wave function of probability collapses.

Now suppose that you do not look at the screen—instead, only your friend Alice, who is in the lab with you, looks. She sees a definite outcome of the experiment, that is, a black spot somewhere on the screen. Relative to Alice, the wave function has collapsed. But relative to you, the wave function remains uncollapsed, that is, it still reflects a superposition of all possible impact spots on the screen. And because Alice has looked, she has become entangled with the particular outcome reflected by the spot on the screen.

What this means is that the world experienced by Alice has changed in several irrevocable ways once Alice has seen the spot. She will have a memory of her observation. She may, if it's a slow news day and nothing much has happened in her life, tell a couple of friends about what she observed and what she thinks it means. They may tell others, and maybe one of these friends then sends a tweet about it to 251 people, five of whom find it important enough that the information changes decisions in their lives. Inspired by the experiment described in the tweet, one of them, Emma, decides to go back to school to study theoretical physics. But on her way to her first class six months later, she gets into a fender bender in the college parking lot. This is how she meets Michael, the other driver and a physics teacher, and while their relationship began with Emma yelling at him for not watching where he was going, they ultimately get married and collaborate to create a key nuclear weapons improvement. That technology is later stolen by a terrorist group preaching a radical anti-hip-hop agenda, and the group detonates their weapon at the Rock and Roll Hall of Fame.

All of these events, including the destruction of Cleveland, are intimately linked to the spot Alice saw on her monitor. These events arise or fail to arise just as the dot arises or fails to arise. Together they constitute a "world" that is either a possibility, or—according to the many-worlds interpretation of quantum theory, first proposed by physicist Hugh Everett in the 1950s—an actuality comprising a sort of alternate reality.

But relative to you, the wave function of the screen and Alice seeing a black dot—along with her life and that of her friends—remains in superposition. This superposition situation contains many versions of Alice, each one seeing the black spot at a different place on the screen or not seeing it at all. When you yourself look at the screen, then you observe a definite spot and hear from Alice that she also sees the spot at the same place. Before the measurement, there were many possibilities, which we'd define as many possible worlds, but after the measurement, your consciousness "hung up" on one of those worlds.

According to the interpretation of QM formulated by Everett, those many worlds are not just hypothetical, they actually exist as components of a universal wave function that evolves like a branching tree and never collapses. Instead of a collapse that ends all possibilities then and there, each measurement causes the wave function to split, with each resulting branch containing a copy of the observer with a distinct memory about a specific observed result (figure 4.4). For instance, in one branch you and Alice see a black spot in the upper left corner of the screen, while in another branch you see it in the lower right corner, and so on. Each branch is a "world" experienced by a copy of you and Alice. From the point of view of each copy of you, the wave function has collapsed,

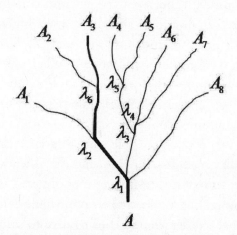

Fig. 4.4 Wave function represented as a branching tree. The bold line represents the path of consciousness. The other paths do not belong to my experiences, but to the experiences of copies of me.

changing from one encompassing the superposition of many possible outcomes of the measurement into a wave function that reflects only one outcome.

The other key point here is that from the perspective of some other observer who has not looked at the screen, the wave function remains uncollapsed and contains many copies of the screen and you. For instance, if Alice does not look at the screen, she perceives a wave function with many copies of the screen and you. Similarly, if you do not look, then you perceive a wave function that embraces many copies of the screen and Alice.

The above examples with you and Alice make it clear that a wave function comprising a restricted set of possibilities *is always relative to some observer*. It is the first and easiest proof that wave function is observer dependent, and it illustrates that this statement is not at all fuzzy or open to interpretation, and certainly no mystical attempt to turn physics into a yoga retreat, as some have tried to imply.

To see this more clearly, take another example of a restricted set of possibilities that changes the wave function: think of what happens if initially the electron is confined within, say, a box. The electron's position is somewhere within the box, and the wave function reflects that. If we had not confined it, however, the electron could be anywhere in the universe. And even now, the value of the wave function can flexibly change depending on our actions: if we open the box, the electron's wave function starts spreading, and after a sufficiently long time, the probability becomes uniformly spread over all the universe. So if we do not confine a configuration of particles, then the wave function includes all possible configurations. Along the same lines, if the wave function is *not* spread uniformly over all possible configurations, this means that it must have been observed, measured, or interfered with by an observer. If so, such a wave function is relative to that observer. There can be no wave function comprising a restricted set of possible configurations that would not have been observed and thus relative to some observer. The same also holds for the wave function of the universe. It represents a universe experienced by that observer, for instance Bob, and contains other observers, such as Alice.

In light of the above reasoning, we want to be sure we don't make the common mistake, when trying to understand Everett's many-worlds theory, of visualizing the "universal wave function" as floating somewhere out there permeating the universe and operating independently. If we did imagine such a thing, we'd also have to imagine ourselves as unneeded bystanders. Instead, we should bear in mind that to embrace all possibilities, all configurations, even all possible universes, particles and objects and energies of all kinds will not manifest themselves unless they've been perceived or in some way meddled with by some observer, and so they are relative to that observer. Thus, no configuration of the cosmos's contents unfolds independently of us. In other words, while the overarching or universal wave function is sometimes imagined as being synonymous with all possible worlds including pre-meddling *and* post-meddling situations, we mustn't let ourselves believe in ghosts: absent observation and its intimate correlation with consciousness, we have only make-believe.

If, dear reader, you feel your hold on all this is still a bit fragile, don't worry. You have lots of company, including many of the physicists who first encountered these ideas. We've seen how these revelations spring directly and indirectly from quantum mechanics, but it's worth mentioning that, in actual practice, physicists exclusively "observe" quantum laws operating through the eyes of math, whereas we are now attempting to do so via clunky verbal descriptions and analogies. Hang in there: We have been laying an important foundation, but it is merely the launchpad for our exploration of How Everything Works. Your investment in fully grasping it will yield reality-shattering dividends.

In this chapter, we've revealed how the actual arises from the possible, and what wave function and observers have to do with all of it. We've also seen how, while we generally dismiss the "multiple universes" of science fiction as just that—fiction—there may be more than a morsel of scientific truth to this popular trope. The alternate realities we hinted at in Chapter 2 may well be more than *what ifs*.

If so—and everything that could possibly happen actually does occur in some Everett universe—then of course, death does not exist in any real sense because consciousness and experience always continue unabated (more about this in Chapter 10). All possible universes exist simultaneously, regardless of what happens in any of them. Thus, in some worlds, Napoleon was not defeated at Waterloo. In some worlds, Alexander the Great was not born. And you, in high school, really did date the homecoming queen or starting quarterback.

DOWN WITH REALISM

5

What do we perceive besides our own Ideas or Sensations?
—**Bishop George Berkeley**

In the never-ending human quest to understand life's underlying realities, a profound challenge arose early in the twentieth century with the advent of quantum theory. Until then, science agreed with common sense, which dictated that each individual's perception of nature was far less important than nature itself. After all, a rock's composition and its location were trustworthy facts, while someone's measurement of such qualities was iffy and subject to revision.

This classic commonsense view came to be known as "realism," and it reflects what most laypeople still believe to be true: the objective world "out there" is real, while each person's "take" on it is tentative. Even our language supports this distinction. We say, "Try to be objective," and "Don't let your subjectivity influence your report."

There is a common saying often invoked by statisticians and scientists about models that are "usefully wrong." In everyday life, the principles that go along with realism have some obvious usefulness: being aware of bias and subjectivity is crucial when evaluating the things people tell us, and no one attempts to walk through a door without opening it (or if they do, they don't make the same mistake again). But let's pause to look at the definition of "realism" as it is meant in physics. In the literature, one finds this: "The doctrine that material objects exist in themselves, apart from the mind's consciousness of them." Another source defines realism as the principle that "matter has its own existence independently of our mind," and yet another definition is "The quality of the universe existing independently of ourselves." One calls realism "the view that reality exists with definite properties even when it is not being observed."

If you've been paying even the most glancing attention over the past few chapters, you see the problem. At the quantum level, the premise that nature's entities have objective properties like motion and position that exist independently of any measurement of them has been flatly contradicted by experimental and observational data. Today, biocentrism's view that nature and the observer are correlative is widespread, and the notion of a definite material universe existing independently of consciousness, though still embraced by the vast majority of the public, is challenged by many in the physics community. While it's a safe bet that your door will be painfully present if you try to walk through it, the location of the individual particles making it up remains a matter of probability. If it helps, Einstein was just as resistant to this as most laypeople are today—without realism, we're living in that dice game he so deplored.

But realism wasn't the only bit of science's shoreline being washed away by the tsunami of quantum theory in the first decades of the twentieth century. Also swiftly disappearing was "locality." This, too, was an age-old commonsensical belief, namely that a thing can only be jostled, moved, or influenced by something in its immediate vicinity in direct contact with it. Thus, everyone knew that a fluttering flag seen outside their window was surely being forced into motion by a nearby material substance, even if that "acting body"—in this case, the wind—couldn't itself be seen.

The erosion of locality began with none other than our old friend Isaac Newton, who, as we saw in Chapter 2, gave us our first trustworthy understandings of how things move, and introduced us to the "force" he named after the Latin word *gravitas*, which reaches out with invisible claws to influence objects big and small alike.

But something about his own gravity descriptions bothered him. It was the locality business, even though that specific term wouldn't be coined for two more centuries.

In letters around 1692, he tortured himself over the notion that he'd put forth a supposed force that made objects shift position without being physically touched as a flag is touched by the wind.

"That one body may act upon another at a distance thro' a vacuum, without the mediation of anything else . . . ," he wrote, "is to me so great an absurdity that I believe no man who has . . . a competent faculty of thinking can ever fall into it."

So Newton, in the sort of anguished introspection rare for the period, expressed his deep fear that his revelations were in some sense impossible, even though his "laws" were consistently borne out as true.

Of course, gravity was just the beginning of physicists' discoveries of invisible forces. Some of these energies operated via "fields," introducing the concept of forces that leaked into their surrounding airy space along the delicate tendrils of unseen pathways. It was all so eerie, and yet it could indeed explain some otherwise baffling behavior of physical objects, for example magnets, which can be controlled via magnetic fields both to display hands-free motion as well as ironclad resistance to movement. Today, such magnetic fields are utilized in automatic gate-locking systems that no reasonable amount of force can thwart.

Einstein, usually the iconoclast, nonetheless maintained an unwavering adherence to the principle of locality all his life. A little over two centuries after Newton's self-doubt, Einstein formulated his relativity theories with strict adherence to locality principles, including an important additional tweak: the effect of any object or bit of energy or field is constrained by a specific travel velocity, the speed of light. In other words, instantaneous influence was impossible.

This meant that in calculating the effects of an event, we could be certain that the fastest consequences we'd ever observe would involve the delay of one second for each 186,282.4 miles of separation. It meant that if a star blew up to be a supernova, earthly sky watchers would see this happen with a lag time that precisely matched the star's distance in light-years. The brilliant glow of a supernova happening now, one hundred light-years away, would first light up our sky in the opening quarter of the twenty-second century.

Einstein insisted that gravity would obey locality, too, so that when two ultradense black holes collided and merged 1.3 billion light-years away, the resulting gravitational ripples that reached the newly built LIGO detectors in Chile's Atacama desert in 2017 had originated 1.3 billion years ago.

With the small distances found here on Earth, cause-and-effect delays are negligible, yet—at least these days—they are measurable. According to the rules of locality and Einstein's light-speed limitation, this amounts to a billionth of a second of delay for each foot of separation. Thus, when you wave to a friend spotted across the street 50 feet away, her response or shouted "hi!" is delayed by 50 nanoseconds or 50 billionths of a second, the time your image takes to reach her eyes, though this invariable hesitation has yet to create an awkward social situation.

What *was* awkward, at least for many physicists, was the advent of quantum mechanics, whose equations bewilderingly insisted that tiny-object effects and influences would be totally free of all such light-speed-induced limitations and pauses.

Remember, this was one of the reasons that quantum theory's predictions about entanglement were so unacceptable to Einstein and his buddies Nathan Rosen and Boris Podolsky, who together wrote the seminal 1935 paper on the subject, as discussed in a previous chapter. According to quantum theory, an entangled particle's "detection" of its twin's state and its own response occurs instantaneously, in real time. The "information" of the initial wave function collapse does not require one billion years to spread across space and reach the twin, even if it's in a galaxy a billion light-years away. Instead, the twin knows and responds instantly. Bye-bye locality. And, since the entangled particle only assumes its definite property in response to the measurement of its twin, entanglement offered more evidence that realism had left the building as well.

But locality and realism had ruled human thought since Neander-thals were running around in furry underwear, and physicists weren't going to let it go without a fight.

In one corner stood Einstein, Rosen, and Podolsky, the famous "EPR," the sheriff's posse guarding the jailhouse of classical physics. These three stated flat-out that if the predicted entanglement effects did indeed happen, they would have to be due to some unknown hidden variable, or some contamination of the experiment. Einstein and Co. continued to believe that nothing could influence anything else unless it made some kind of direct contact, even if only through the medium of energy fields, that light speed was an absolute velocity limit, and—most importantly of all—that it all happened whether or not we were watching. (Still, Einstein was known to have privately asked a colleague, "Do *you* believe the moon exists when no one's looking?") They defended the commonsense view with their own reputations, because they believed heart and soul that, come what may, locality and realism would have to be maintained.

Meanwhile, it was 1935, and the Depression-era public was struggling to afford shoes. And, soon after, dealing with the Axis powers trying to take over the world. Few were aware that an epic battle was raging in academia over whether the Messerschmitts existed when nobody was looking at them.

The issue still hadn't been fully settled when your authors went to school. Then, in 1964, came the theoretical work of John Bell, which used probability to explore the likelihood of different measurements of entangled objects based on various detector settings. It is a mathematical proof too complicated to go into here, but the upshot is that the probabilities don't match what would be expected if some hidden local variable was the source of the strange behavior of entanglement. Experiments over the next twenty years, particularly by Alain Aspect, dealt fatal blows to Einstein, Podolsky, and Rosen's local variable stance. However, these experiments had loopholes, mostly due to the limitations of testing equipment, and it wasn't until the years immediately surrounding the birth of the twenty-first century that actual laboratory experiments were able to conclusively demonstrate the defeat of local realism.

The key initial proof, the 1997 experiment by Nicolas Gisin discussed in a previous chapter, used equipment able to measure delays thousands of times shorter than the twenty-six-thousandths of a second that light would have required to convey the information across the seven-mile distance separating the entangled twins in Gisin's Swiss laboratory setup. This newly certified fact of faster-than-light information transfer was a game changer—the speed of light speed limit was dead.* But science geeks still wondered: Does the entanglement business happen at merely faster-than-light speeds, or was it truly instantaneous? Breaking the light-speed barrier was enough to shoot down EPR's defense of classical physics, but travel in zero time carried unprecedented implications about the very nature of reality.

You know researchers—if there's something to be quantified, they'll do anything to pin that number down. So in 2013, Chinese physicists entangled pairs of photons, then transmitted half of the pair to receiving units that were positioned nearly ten miles apart in an east-west orientation designed to minimize confounding effects caused by the 1,038 mph speed of Earth's rotation. Essentially, the task was to measure one member of the entangled pair and then see how quickly the other assumed a complementary state, being sure to repeat the procedure often enough, for a full twelve hours consisting of numerous measurements, to account for any outliers and narrow in on precisely how much time elapsed between the measurement of one photon and the response of the other.

At day's end, the Chinese team found that quantum entanglement exchanges information at around 20 million miles per second, or about 10,000 times faster than the speed of light. That mind-twisting speed, equal to 4,000 Earth-to-moon round trips in one second, made headlines. Nonetheless, while it supports the "instantaneous" prediction of quantum theory, it is not conclusive. Since that was the lower limit of their experimental testing abilities, the real figure will almost certainly be faster. And the "no time at all" prediction of quantum theory remains

---

* The information considered here is about the quantum state of the entangled twin. It is not possible for two persons to communicate by exchanging messages faster than light by employing quantum entanglement.

safe, having aced every experimental test thrown at it. Indeed, experiments showing both violations of locality (instantaneous effects from distant hands-off sources) and violations of realism (the idea that objects exist with definite properties even when they're not being observed) are nowadays announced regularly.

With experimental proof of QT's predictions pouring in, and the local realism castle having come crashing down for good, scientists, philosophers, and metaphysicians find themselves strange bedfellows, temporarily joining forces to wonder what it all implies on deeper, meaning-of-life levels.

Certainly some sort of interconnectedness across the universe has now been established. Some as-yet-unknown property of non-separation between objects, no matter the supposed distance between them. A belief in "oneness" is no longer solely the province of mystical types. "Non-separability," said physicist Bernard d'Espagnat, "is now one of the most certain general concepts in physics."

This was an important bridge to the biocentric model. And, too, one more nail in the coffins of space and time. Sure, those concepts still serve a useful purpose melded into the mathematical amalgam of Einstein's "spacetime," which lets us calculate how classical objects must move through the universe, and how observers in one "reference frame" of speed and gravitational force observe objects and events in another—we'll talk more about spacetime in a later chapter. But "time" and "space" as reliable and distinct components of some external matrix? Well, any pretense at an inherent stand-alone reality for those two is now gone for keeps. (Ironically, Einstein himself was the first to pull the rug from under them, by showing in his relativity theories that both space and time can warp, shrink, and even collapse to nothing depending on local circumstances.)

But if time and space are no longer trustworthy, independent entities, how then should we visualize our universe and our place in it? How then—and where—should we picture events unfolding?

Whenever the term "solipsism" arises, it's invariably offered as a dangerous endpoint to be avoided in any scientific discussion. Yet it's not difficult to understand why the word periodically crops up in the first place. A thorough examination of the implications of quantum theory is an excursion along a trail skirting solipsism's slippery boundaries.

The *Random House Webster's Dictionary* defines solipsism as "the belief that only the self exists or can be proved to exist." When most people encounter this for the first time, the reaction is usually, "That's ridiculous." But such a facile dismissal quickly fades as one looks more closely.

Everyone's heard the most famous pronouncement of René Descartes: *Cogito, ergo sum*—"I think therefore I am." Less well known is another quote of his: "The first precept was never to accept a thing as true until I knew it as such without a single doubt." Descartes was obsessed with being sure that the evidence he used to construct his worldview was reliable.

His was a basic question, one that serves as the foundation for any inquiry into the nature of reality: *Of what, indeed, can any of us be totally sure?* Throughout history, all sorts of seemingly ironclad statements about reality have been marched before the public and presented as absolute truths, but always there are loopholes or inconsistencies. In seventeenth-century France, Descartes was surrounded by a community of thinkers who were intent upon delineating an objective, matter-based universe, removing the subjective observer from the equation. Yet, despite being embedded in this milieu, Descartes realized that he could never be totally certain about the nature of the material cosmos. For how could he be certain, he reasoned, that everything he perceived was not merely in his mind? Rather, he could only fully depend on the fact of his own experience.

Descartes was hardly alone in considering such lines of reasoning, nor was he the last to reach the same conclusions. In the eighteenth century, Bishop George Berkeley had similar revelations. "The only things we perceive are our perceptions," he famously said, asserting that we merely *assume* such perceptions correspond to an actual world of objects external to us. By denying the independent existence of material substances, or at least any certainty about them, Berkeley broke completely from the rational philosophy popular in the period, and he infuriated a whole slew of his contemporaries.

Of course, not being able to be certain something is real isn't the same as proving that it isn't. But with the rise of QT and experiments showing that, at least in the realm of the quantum, material objects don't exist with definite properties before we observe them, the idea that the universe isn't an objective external reality is suddenly supported by science, not just philosophers following logic down a rabbit hole. And, as Heinz Pagels, an esteemed theoretical physicist, once said: "If you deny the objectivity of the world, unless you observe it and are conscious of it (as many physicists have), then you end up with solipsism—the belief that your consciousness is the only one."

Pagels's conclusion was right. Only, according to biocentrism, it isn't your consciousness that is the only one, it is *ours*. Our individual separateness is an illusion. After all, if space and time do not exist in any absolute sense, then in what way can we think of things as being separate? There is one single consciousness. What is "focal" (what you experience as yourself) is this single consciousness manifesting itself in various different ways.

So is this really solipsism, or its opposite? Solipsism and the belief in universal oneness—"only self" and "no self" aren't as easy to separate as you might think. After all, in a way, one leads to another. They're like the twisted strands of a single piece of thread.

As we've noted earlier, this kind of thinking already enjoyed a venerable pedigree by the time it was tossed around in Renaissance Europe. Way back in the sixth century BC, the philosopher Parmenides concluded that a single deathless essence was the nature of the universe. And that this cosmos, which was identical with our consciousness and in no way separate from ourselves, had neither a birth nor would it ever perish. It was also immune to change, at least on a fundamental level. In his poem "On Nature," he affirmed the absolute primacy of consciousness, writing (centuries before Descartes), "To be aware and to be are the same."

And that still wasn't where the one-mind business originated. Even earlier, Shankara and other Hindu writers had insisted that "all is one," and that this oneness was identical with the self. A bit later, various branches of Buddhist philosophy jumped on the bandwagon, most notably Zen Buddhism. They claimed that the so-called experience of enlightenment boiled down to the direct perception of oneness. Having this specific experience became the core goal for practitioners of Eastern religions and remains so today both among adherents of those religions and those who are devotees of the increasingly common practice of meditation. In this perception of "truth," the meditator perceives that, in reality, there is no "self"—and there are no "others."

And for a relatively modern mathematician who agrees with the meditators, you need look no further than Schrödinger, who was both one of the founders of quantum theory and, as we'll discuss in the next chapter, one of its harshest critics. He was also way ahead of the pack when it came to the connection between quantum theory and consciousness, perceiving before most a fundamental connection between the basic physics of the universe and the foundations of perceptual reality: "Every conscious mind that has ever said or felt 'I' is [the one who] controls the 'motion of the atoms.'"

He was also ahead of the game when it came to non-separability. One problem from the outset was that an "all-is-one" paradigm seemed very much to contradict the evidence of our everyday experience of separate consciousnesses. After all, my dreams are not the same as yours, nor can I wiggle your toes. This commonsense view had nearly universal support in the Western model that took for granted innumerable points of control, at least separate bodily control, which in turn implied multiple islands of independent consciousness.

"That's a delusion," insisted Schrödinger in writings that would have provoked applause from the priests of Varanasi. "Even what seems to be a plurality is merely a series of different personality aspects of this one thing, produced by a deception."

He went on to explain, "The plurality that we experience is only an appearance; it is not real . . . I should say: the overall number of minds is just one. I venture to call it indestructible since it has a peculiar timetable, namely mind is always now. There is really no before and after form. There is only a now that includes memories and expectations."

On other occasions, he liked to say, "Consciousness is a singular of which the plural is unknown." And bingo, here we are back to solipsism again, or at least its intersection with the idea of an overarching oneness.

Interestingly, without using the word specifically, more and more science fiction titillates audiences with solipsistic story lines. In the wildly popular *Matrix* movies, the protagonist Neo is presented as essentially a "brain in a vat" whose seeming adventures within a vast external world are in fact artificially produced and actually occurring strictly in the mind. (In that case, yet another external world, of which humans are largely unaware, holds them captive.)

In the holographic universe model increasingly put forth in popular science magazines, nature is explained as an artificial design akin to a hologram on a bank card, where perceived colors, dimensionality, and the presence of multiple people and other living organisms who come complete with complex, interwoven story lines are no more than computer code.

In this context, one-mind or consciousness-based models may no longer seem particularly far out. Certainly, the overall direction of the sciences over the past 150 years has been toward a search for unifying explanations, with the discovery of numerous simplifications as a result. From the nineteenth century, when electricity and magnetism were found to be aspects of a single phenomenon, the Pandora's box of unification in the sciences has been wide open, beckoning others to follow. In the early twentieth century, Einstein united first matter and energy and then space and time, and later, midtwentieth-century theoreticians seeking to uncover the conditions that existed in the minutes and seconds following the big bang discovered that three of the universe's four fundamental forces had originally been merged rather than existing as separate entities. And today, many physicists no longer speak of the "weak force" and the "electromagnetic force" as distinct, but instead deal with what they call the "electroweak."

The point is, it is undeniable that the underlying unity implied by science's quantum mechanical breakthroughs has been and remains a core quest whose endpoint has yet to be fully realized. It is the job of science to prove or disprove the role of the observer, and to consider the implications of oneness suggested by the experimental results of the past 150 years.

And science must do all this without flinching, following the evidence even when it contradicts our longest-held and most fundamental beliefs.

Standing in the ruins of local realism, as evidence accrues of not only a true interconnectedness but an interconnectedness that explicitly involves mind or consciousness, we're finding ours to be a cosmos as simple and united as can be imagined—and one that shares its most intimate identity with our very selves.

CONSICIOUSNESS

I regard consciousness as fundamental.
—**Max Planck**

Twentieth-century physicists were taken aback by their sudden awareness of "awareness," that most fundamental aspect of human existence.

On the one hand, awareness or consciousness had an inarguable
reality; more so, perhaps, than even the most solid math-based conclusions about the material universe, the focal point of their studies. On the
other hand, consciousness seemed out of place in a science conversation;
it was sort of like discussing love, relationships, or other such imponderables. That was partially because scientists, led by the likes of Pierre
LaPlace in the prior century, had more or less succeeded in making the
universe seem like a vast self-operating machine. Discover the laws of
motion, the rules of probability, the nature of the forces that pull and
push things, and you could predict everything in the mechanical cosmos.
Awareness need never come into it.

Yet advancements in physics in the 1920s kept bringing the observer
and awareness front and center. And the brilliant minds that collectively created and refined quantum mechanics—Max Planck, Werner

Heisenberg, Niels Bohr, Erwin Schrödinger, Wolfgang Pauli, Albert Einstein, Paul Dirac, and, later, Eugene Wigner, among others—came to realize that there was a big stumbling block to a strictly objective model with all contaminating human observers removed from the picture. As noted by Heisenberg, it was this: "The transition from the possible to the actual takes place during the act of observation. If we want to describe what happens in an atomic event, we have to realize that the word 'happens' can apply only to the observation."

Quantum theory kept showing that at any particular moment an object like an electron or photon could be a wave or a particle but not both, or could have an up spin or a down spin, or a horizontal or a vertical polarization, or could be here rather than there, and which properties it would be observed to have could never be predicted ahead of time. The process of materializing or appearing in one way rather than another involved an instantaneous change in the object's wave function, which, as we've seen, was a strange preexistence as a kind of blurry potential or probability, one that had not yet "collapsed" into an actual item with tangible properties. The question of the day was: What causes this wave function to collapse and give birth to the object as an actual enduring entity? And, according to studies like the famous double-slit setup, the defining factor seemed to be the observer, someone making a measurement.

In the fullness of time, the role of the observer has proven more rather than less central than first imagined. Not only do the actual hard-core properties of reality, such as whether an electron shows itself as a wave rather than a particle, mutate depending on the presence or absence of information in an onlooker's mind, but, prior to an observation, it doesn't even make sense to speak of photons or subatomic particles as possessing attributes to begin with! Indeed, these days it is fully mainstream physics to say that an electron does not have any real position in space or any actual motion independent of the observer.

As the great Princeton physicist John Wheeler once declared: "No phenomenon is a real phenomenon unless it is an observed phenomenon." Meaning that the word "observation," despite seeming to imply a passive process of being an onlooker, is actually the practice of reality creation.

Thus, roughly a century ago, as experimenters were first beginning to show that the so-called external world physically changed depending on our observations, Heisenberg wrote, "The discontinuous change in the wave function takes place with the act of registration of the result by the mind of the observer. It is this discontinuous change of our knowledge in the instant of registration that has its image in the discontinuous change of the probability function."

And, as Heisenberg went on to explain, "The observer is never entirely replaced by instruments; for if he were, he could obviously obtain no knowledge whatsoever. The instruments must be read. The observer's senses have to step in eventually. The most careful record, when not inspected, tells us nothing."

In short, as we said earlier, all observations (even measurements made by instruments) can only be known to us by consciousness, and so thanks to the newly discovered role of observation, consciousness became an unexpected focal point for serious probes of physics. It was where students of natural laws had to turn—to this phenomenon that now was clearly perceived not only to apprehend the cosmos, and not only to physically alter its contents, but as the mechanism that was responsible for it manifesting in any way.

So, around the end of the First World War, the world's greatest physicists were suddenly talking about what previously had been a locked-in-the-attic entity dealt with only by metaphysicians, philosophers, the clergy, and mystics. Diving into this mysterious realm must have been both weird and frustrating as consciousness has long proved to be a slippery subject, one that is extremely resistant to examination by the usual scientific methods.

Nonetheless, it seemed every early twentieth-century theoretical physicist was joining the choir in praise of the awareness motif:

"Everything we call real," said Bohr, "is made of things that cannot be regarded as real. A physicist is just an atom's way of looking at itself."

Said Pauli, "We do not assume any longer [the reality of a] *detached observer*, but one who by his indeterminable effects creates a new situation, a new state of the observed system."

The consciousness infection was spreading to them all. Even the most equations-obsessed founders of quantum mechanics saw that their newly effective way of probing the submicroscopic realm forced them to contemplate the observer himself:

"I regard consciousness as fundamental," wrote Max Planck, with a confident open-and-shut tone akin to that of the Sermon on the Mount. "I regard matter as derivative from consciousness."

And lest we imagine that those early quantum physicists were simply victims of a kind of consciousness craze sweeping postwar Europe, quantum geniuses later in the century kept up the same chorus.

As Nobel Prize–winning Hungarian American physicist Eugene Wigner explained in 1961, "Until not many years ago, the 'existence' of a mind or soul would have been passionately denied by most physical scientists. The brilliant successes of mechanistic and . . . macroscopic physics overshadowed the obvious fact that thoughts, desires, and emotions are not made of matter, and it was nearly universally accepted among physicists that there is nothing besides matter. The epitome of this belief was the conviction that, if we knew the positions and velocities of all atoms at one instant of time, we could compute the fate of the universe for all future . . . [But after the advent of quantum theory] the concept of consciousness came to the fore again: It was not possible to formulate the laws of quantum mechanics in a fully consistent way without reference to consciousness."

He later summarized it this way: "The very study of the external world [leads] to the conclusion that the content of consciousness is an ultimate reality."

John Bell, the Northern Irish physicist whose theorem famously provided a mathematical basis for locality-violating entanglement a few years later, echoed Wigner, declaring, "As regards mind, I am fully convinced that it has a central place in the ultimate nature of reality."

As we'll see later, physicists from Hawking to Wheeler have taken things even further in the decades since, with concepts like the "participatory universe" in which we don't merely create the present but the past as well. As the famed British cosmologist and Astronomer Royal Martin Rees has said, "The universe could only come into existence if someone observed it. It does not matter that the observers turned up several billion years later. The universe exists because we are aware of it."

Well, far-out indeed! But for now, the point is that, starting around a century ago, physics took an abrupt turn and began to seriously consider that without consciousness, the material universe alone could not supply the true or complete picture of reality.

The most famous early rebuttal to this "observation changes reality" bottom line came from Erwin Schrödinger, who, despite fervently believing that an eternal all-is-one consciousness pervades the cosmos, objected to what he saw as the illogical conclusions of quantum theory as outlined in the Copenhagen interpretation.

Again, to quickly review, the Copenhagen interpretation, named for famous Dane Niels Bohr, was the consensus interpretation of QT as we've explored in previous chapters. It held that a quantum system—an atom, say, along with any observers who might be watching or affected by it—will decisively assume one state or another only upon observation. Until that observation, all possibilities continue to coexist and are equally real. In other words, a particle might be in two places at once, or a photon might possess both a horizontal and a vertical polarization, and this remains so until someone takes a look. Then one state materializes and the other vanishes without a trace.

The Copenhagen gang dealt with the fact that this kind of behavior didn't make sense in the classical world (think of that example from an earlier chapter in which a baseball is fair or foul but not both) by saying there was one set of rules for the quantum and one for the classical and ne'er the twain shall meet. To poke a needle into that whole balloon, Schrödinger set up a hypothetical situation showing that the two worlds could be connected, producing what he said would be "a ridiculous case." In 1935, in the German publication *Naturwissenschaften*, he wrote, "A cat is penned up in a steel chamber, along with the following device (which must be secured against direct interference by the cat): In a Geiger counter, there is a tiny bit of radioactive substance, so small, that perhaps in the course of the hour one of the atoms decays, but also, with equal probability, perhaps none; if it happens, the counter tube discharges and through a relay releases a hammer that shatters a small flask of hydrocyanic acid. If one has left the entire system to itself for an hour, one would say that the cat still lives if meanwhile no atom has decayed. The first atomic decay would have poisoned it. The psi-function of the entire system would express this by having in it the living and dead cat mixed or smeared out in equal parts."

In short, quantum theory said the radioactive atom in the box would exist in superposition before it is observed, which would mean that the hapless cat was simultaneously dead and alive until the box was opened—something everyone agreed was impossible. (At least in a single world . . . but Everett's branches were yet to come!) Schrödinger's point was that the Copenhagen interpretation seemed to make this absurd conclusion inevitable, and so must be flawed.

This Schrödinger's Cat business became the most famous thought experiment in history, but it was not exactly original. The first time a physicist illustrated the illogic of this QT interpretation by finding a way to entangle submicroscopic quantum behavior with our visible everyday classical world came a whopping fifteen years earlier in 1920, when Albert Einstein concocted a very similar thought experiment using the example of an exploding bomb similarly triggered by atomic decay.

Now, there's no arguing with Schrödinger or Einstein on one point: the prediction of quantum theory that reality depends on an observer is deeply weird. But the fact is, experiments continue to bear it out.

A bit over half a century ago, in 1961, Eugene Wigner outlined yet another famous thought experiment. It involved two observers—Wigner and Wigner's friend—one making a measurement inside a lab, and another hearing about it afterward, much like the situation discussed in Chapter 4. It was designed to explore the nature of measurement and whether objective facts can exist. If the state of an object remains in superposition for the observer outside the lab until they are told the result, but "collapses" upon measurement for the observer inside the lab, what does that mean for reality and the role of the observer in it? Hypotheticals like this reflect the long suspicion among physicists that quantum mechanics allows two observers to experience different, conflicting realities, but recent advances in quantum technologies have at last made it possible to test this in a lab, using entanglement. In a state-of-the-art experiment that was published in *Science Advances* in 2019, Massimiliano Proietti and his colleagues at Heriot-Watt University in Edinburgh created different realities (using six entangled photons to create two alternate realities) and compared them (see figure 6.1).

Fig. 6.1 Experimental setup of Science Advances *paper published in 2019 (Massimiliano Proietti et al., Sci Adv 2019;5:eaaw9832). Pairs of entangled photons from the source (SO) were used to create two alternative realities, which were distributed to Alice's friend (left-hand box) and to Bob's friend (right-hand box), who measure their respective photons to determine if the measurement and the photon are in a superposition.**

Despite using state-of-the-art quantum technology, it took the researchers several weeks to collect enough data to have statistical power. But eventually, the experiment produced an unambiguous result—that both realities can coexist even though they produce irreconcilable outcomes, just as Wigner predicted. Realities can be made incompatible so that it is impossible to agree on objective facts about an experiment. These results suggest that objective reality does not exist. Wrote the authors: "This result implies that quantum theory should be interpreted in an observer-dependent way . . . The scientific method relies on facts, established through repeated measurements and agreed upon universally, independently of who observed them. And yet in [our] paper, they [the observers] undermine this idea, perhaps fatally."

Although the observers in the experiment were modeled by entangled photons,[†] the same principle also applies to macroscopic detectors, including conscious observers. The results provide experimental confirmation of what we've explained in this chapter, and what scientists last century first realized, that what an observer is aware of—in other words, consciousness—changes reality.

So consciousness is important (perhaps the understatement of the century), but studying it has always brimmed with unpleasant pitfalls. What exactly *is* it, anyway? Despite some ongoing controversy over its definition, it's generally agreed to mean the state of being aware, of perceiving things, of having feelings, of wakefulness, of possessing experiences. The slipperiest aspect of pinning down consciousness is that understanding it means not just probing the nature or quality of thoughts, but, more to the core, what does *it feel like* to have thoughts? In recent times, the word "qualia" has been used to refer to these individual, subjective sensations or experiences that define consciousness.

The philosopher David Chalmers coined the phrase "The hard problem of consciousness" to denote the difficulty science has in trying to explain how matter—carbon, hydrogen, and oxygen atoms, or brain tissue, or moving electrons (electric current) traveling along neurons—gives rise to the subjective experience of a purple twilight sky or the smell of freshly cut grass. To date, all explanatory efforts have proven futile. And, even harder to explain is how *any* qualia, these sensations of perception, can arise at all in the first place. It resurrects centuries-old debates over whether matter and consciousness are distinct, or are linked—and, if they are separate, which is the more fundamental.

At times, the whole subject feels like an elaborate tease. Awareness is probably the most intimate and obvious aspect of reality, which makes it extremely ironic that it remains impossible to explain and challenging even to discuss. Exploring it manages to be simultaneously stone simple and unachievably difficult. This particular torment stems from the

---

† This will be further clarified in the next chapter in which we will introduce the concept of the hierarchy of representations, according to which, from the first person's point of view, other observers—just like detectors—are sort of pictures ("representations") in consciousness.

contradiction between our inability to account for how whatever it is that gives animals a sense of awareness arises and the fact that qualia are self-evident to the point of being ineffable. Consciousness lets us perceive the sky as blue, an experience that, while simple and inarguable, couldn't possibly be grasped if we'd been born blind even if someone were to spend endless time trying to convey it to us intellectually. The blue experience is self-evident and unlike anything else. It's also gratifyingly complete. When seeing the sky, we *fully* know what it looks like, with nothing more to attain. Perception is thus comprehensive. It lacks nothing.

Thus, if one is setting out to explore the universe, to probe its characteristics through knowledge, the fact of awareness might very well be viewed as the initial, surest facet of existence. If there is a cornerstone, a starting point, a foundation block, it is consciousness.

Yet, despite early quantum scientists having agreed upon and demonstrated its importance, many scientists see an oil-and-water incompatibility in attempts to throw mathematics and physics equations at life's classical imponderables. And so today, a century later, a majority of scientists change the subject whenever consciousness is brought up, and those who do engage with it persist in doing so superficially—perhaps because, as mentioned earlier in this book, they are either unable or unwilling to stretch the bounds of science to accommodate this kind of inherently subjective phenomena.

Take cognitive scientist Daniel Dennett, author of *Consciousness Explained*. He pooh-poohs qualia as even being a helpful concept, and his book, despite the promising title, essentially ignored the "hard problem" and instead spent hundreds of pages describing which parts of the brain control specific functions such as vision—which is why many critics dismissed the work as "consciousness ignored."

The purported effects of consciousness, along with the larger issue of whether physics should deal with it or leave the topic to philosophers and metaphysicians, remains an ongoing controversy. To most physicists today, the topic of consciousness belongs in the same bin as ghosts, God, or the afterlife.

However, there is serious ongoing pushback against those who want to keep physics isolated from life's larger issues. For example, in 2018,

Italian theoretical physicist Carlo Rovelli wrote about the *necessity* for physics to tackle the deepest unanswered questions, even when they may seem philosophical. In a *Scientific American* blog post, he wrote: "Here is a list of topics currently discussed in theoretical physics: What is space? What is time? Is the world deterministic? Do we need to take the observer into account to describe nature?"

The consciousness issue isn't going away. The vast armada of weighty foundational questions set afloat by the pioneers of quantum theory still sail today, with their hulls as deeply hidden in the seas as they were then.

However, at last, in some of the choppiest waters, sunny harbors seem to beckon—just ahead.

HOW CONSCIOUSNESS WORKS

7

The much-discussed question of the communion between the thinking and the extended . . . comes then simply to this: how in a thinking subject outer intuition, namely, that of space, with its filling in of shape and motion, is possible. And this is a question which no man can possibly answer.
—Immanuel Kant

Y ou keep staring at the repair man. His words are starting to sink in. The fabulous and expensive generator you bought a few years ago to keep the lights burning during storms and power failures needs a major repair.

"A head gasket?"

You echo the phrase he just used, fearing that it has a pricey ring to it. "What exactly *is* a head gasket?"

You listen with interest while the mechanic explains the basics of four-stroke engines and why the two big sections of the engine block require a compressive layer that prevents internal gasses and oil from leaking out.

Modern engineering is indeed a marvel. But the real marvel is how your experience of even the very mundane reality of this repairman's visit can be occurring in the first place. How is it that you can perceive this person in front of you in such three-dimensional detail, his words (at least mostly) comprehensible, each of you perceiving events subjectively, while also managing to communicate within a seemingly very real shared reality? How does your consciousness work?

We've seen that the question of what consciousness *is*, its ultimate origin, is largely a nonstarter. That's because consciousness encompasses all of reality—the two are essentially synonyms—so the question amounts to wanting to know the origin of everything. Making it even more fundamentally hopeless, time simply does not exist as an independent item dwelling outside of consciousness, so there is no exterior matrix out of which consciousness/reality could emerge, and from which we could study it.

Ah, but how consciousness *works* is something else entirely. Happily, we've reached the part of the consciousness tangle where scientists can actually provide answers, for "processes" are exactly the kind of inquiries our minds (and the tools of science) can effectively tackle. The workings of consciousness are still rather more complex than those of a head gasket, because the classical science that governs the workings of a four-stroke engine cannot tackle quantum phenomena such as superpositions, where multiple outcomes hang in the air until a wave function collapse makes the entire ensemble work together to produce a single perceived outcome. And consciousness is, as it turns out, a quantum phenomenon.

Let's start our exploration of the "how" of consciousness by coming to a stop at a traffic light. We all agree the stoplight is "red," even though we can never prove that the exact visual experience I call "red" is the same as yours. It doesn't matter because, whatever it is, it stays consistent, and it has since someone thought to name the colors in the first place.

One of the big puzzles of awareness, of course, is how and why we experience something called "red" to begin with. To understand the problem, consider the fact that light is part of the electromagnetic spectrum, which is a continuous gradient of electromagnetic radiation running from shorter to longer wavelengths. Thus, we might experience the visual spectrum as a gradient of brightness—as a continuum of grays ranging from dark to light. It could be a simple quantitative experience. But, for humans and some other animals, it isn't. Instead we have a unique *qualitative* experience. Why is it that, when light falls within very specific ranges of the visual spectrum, we subjectively experience a distinct sensation we call "red" versus, say, "green"?

In 1965, researchers discovered three types of cone-shaped cells in the eye that, when stimulated, are subsequently associated with the visual sensations of red, green, and blue. Stimulation of each type of cone is associated with a unique experience. But how and why? A clue comes from the fact that fully two-thirds of these cone-shaped cells are the so-called "L type" responsible for the sensation of red. This lopsided majority suggests, from the outset, that perceiving light in the "red" range of the visual spectrum is of higher priority than perceiving other wavelengths of light—and thus that our perception of colors serves a purpose.

In evolutionary terms, red likely gets extra attention from the brain because it's associated with alarming, important events like injury, fire, and blood. In life, the sudden presence of that color in your consciousness usually meant either that your bicycle had gone off the road into a field of begonias, or, more worrisome—and, in the early days of humankind, more likely—that blood was pouring down your arm, requiring immediate attention.

This possibility of a life-threatening situation made red the traditional signal of bad news that shouldn't be ignored. We know this instinctively, which is why no one except a contrarian teenager would dream of painting their bedroom a bright red, at least not if they valued a tranquil environment. This explains why red was universally agreed on as the color for things like warning notices and railroad and, later, automobile stop signals. And why even culturally distinct nations and those antagonistic enough toward the West to want to thumb their noses at new modern conventions didn't buck this rule. Obviously, the qualitatively attention-getting experience we call "red" is associated with a deep built-in pattern of emotions and neural connections.

A similarly distinct circuitry comprising labyrinthine clusters of cells is connected with the other colors and cones—each associated with separate areas of the brain. When these cell architectures are stimulated via their respective cones in the retina, we have distinctive experiences: blue evokes the vastness of the sky and yields a much calmer feeling than red, and green conveys countless bygone centuries of plants and vegetation and is a comforting invocation of life.

We believe that these three most basic colors and their various combinations must have had unique survival value during early evolution, and thus they are associated with their own functional pathways in the brain. When the complex relational logic associated with these distinct clusters of cells is brought into the actively entangled region of the brain associated with consciousness, we have discrete sensations even if we rarely give a second thought to the components that make up each of these colors, any more than we can discern the ingredients in mayonnaise or a piece of Cap'n Crunch.

This is but a brief sample of the workings of processes acting below our conscious perceptions and decision-making. To understand those of which we *are* aware, we must return to the cloud of quantum activity that surrounds the brain's countless neuro-electric occurrences.

No one could be blamed for desiring a fuller explanation of what exactly pulls the trigger in the collapse of the wave function. If it is an observation made in consciousness, then why shouldn't a subconscious event count, such as when we suddenly find ourselves in a tense mood but are unaware that it's due to the odd red color of the walls of the club we've just entered? After all, the subconscious is often the decisive factor in such events, as it is with many reflex actions.

The answer is that activities at a subconscious level are in a quantum superposition—meaning, all possibilities simultaneously coexist. But the moment their results pop into reality and conscious awareness, a perceptible "choice" is made. This is key because there are always many possible chains of brain activities (in many possible Everett branches). But when consciousness hangs up on one of them—subjectively perceived as the awareness of a definite outcome—this can now be mathematically described as a collapse of the wave function.

It might be helpful to recall last chapter's summary of Schrödinger's thought experiment involving the most famous cat in the history of physics. In that example, a chain of events began with a radiation source monitored by a Geiger counter. The wave function of the radioactive material was a superposition of two states—one in which there is a decay and one in which there isn't. Let's simplify by transferring this situation to a modern lab, and omitting any possible cat-euthanization and the subsequent hassles of PETA involvement. If there is a decay, the counter detects a high-energy photon and produces a brief click that enters the ears of the lab technician. There the sound, itself just a transient air pressure wave, is transformed into an electrochemical signal that is transmitted via the nerves to the brain, where the processing of the information begins, first at the subconscious level. Then the information is construed in consciousness as "a click of the Geiger counter," followed by a cascade of interpretative judgments in the cerebral cortex. This entire sequence of events comprises one possible chain of brain activity, but note that the strictly physical radioactive decay and the neural responses are all inexorably linked in a single outcome! The other chain corresponds to the case in which there was no decay, and that corresponds to a completely different chain of brain activity leading to the awareness in consciousness that the counter has produced no click. There are thus two possibility branches—one ending with the conscious awareness of a click and the other in which there was only silence—and according to quantum theory, both of these were equally real (in superposition) until the moment of perception. But from my first-person's perspective, I cannot be in a superposition of these two states of awareness, for they are mutually exclusive: Obviously I cannot both hear a click and also *not* hear it. So I find myself in exactly one of those two states of awareness.

Wave function collapse is thus indeed triggered by my perception of one thing or the other. But what may be news to the reader is that the two branches extend to include the radioactive radium, the instrument, its oscillating speaker, the ear's vibrating tympanic membrane, and all those countless brain neurons. All are inexorably a part of a single Everett branch and are inseparable.

How different parts of the brain are involved in a superposition and its collapse into a singular experience depends on details of how the brain processes information, so here is where we must get a little bit technical. All of the brain's neurons process information through electrical and chemical signals. Neurons are electrically excitable, maintaining differences in voltage across their membranes by means of "ion pumps." The ions in the brain are atoms of sodium, potassium, chlorine, and calcium that are missing electrons, which gives them each a bit of electrical charge. They flow along ion channels embedded in the cell's membrane, which generates intracellular-versus-extracellular ion concentration differences. Changes in the cross-membrane voltage can alter the function of these electrically dependent ion channels. If the voltage changes by a large enough amount, an all-or-nothing electrochemical pulse called an *action potential* (also known as a "nerve impulse" or "spike") is generated, which zooms along the cell's axon at anywhere from 70 to 250 miles per hour, to where it can activate synaptic connections with other cells. Thus, all information in the brain is ultimately mediated through ion dynamics.

These ions, as well as the channels through which they enter or leave the cell, are very small. As the American mathematical physicist Henry Stapp has pointed out: "This creates, in accordance with the Heisenberg uncertainty principle, a correspondingly large uncertainty in the direction of the motion of the ion. That means that the quantum wave packet that describes the location of the ion spreads out during its travel from ion channel to trigger site, to a size much larger than the trigger site. That means that the issue of whether or not the calcium ion (in combination with other calcium ions) produces an exocytosis (leaves the cell) is a quantum question basically similar to the question of whether or not a quantum particle passes through one or the other slit of a double-slit experiment. According to quantum theory the answer is 'both.'"

Although Stapp focuses on whether calcium ion channels open or close, there is much more to the mechanism than that. For instance, electrophysiology probes allow us to study the movement of various different types of ions within the cells of the brain. If an electrode is small enough, meaning micrometers in diameter, then it is possible to directly observe and record the intracellular electrical activity within individual cells. Thus, we have the ability to capture the entire mechanism involved in the emergence of time—starting from the quantum level (where everything is still in superposition) to the macroscopic events occurring in the brain's neurocircuitry (see Chapter 11 for more about the brain and the emergence of time).

Talking about calcium channels opening and closing is insufficient, as the equation reduces to merely a cloud of quantum information when you expand the mechanism to include the ion dynamics involved in the whole temporal sequence of events, from changes in ion gradients within the cell to axon firing. And while on one hand the appropriate probes and current technology allow us to monitor the generation and movement of the action potential along the cells' axons, the underlying main story involves the quantum information that arises all at once when the process is expanded to include the ion dynamics and their superpositions.

That's because it is modulation of the ion dynamics at the quantum level that allows all parts of the information system that we associate with consciousness—with the unitary "me" feeling—to be *simultaneously interconnected.*

This is the key. What is relevant here (and for the whole book, whenever we talk about consciousness and the wave function) is that those entangled regions of the brain, which together constitute the system perceived as consciousness in all its manifestations, arise as such because a sense of "time"—or the sequential flowing of events—emerges simultaneously throughout all of the spatial algorithms/neurocircuitry responsible for generating a conscious, real-life (spatiotemporal) experience.

It is important to note that the spatial separation between neurons in the brain is meaningless before this process occurs. It's an all-or-nothing phenomenon.

At any given moment, there is a cloud of quantum activity associated with consciousness. The exact things you feel and consciously experience will change depending on which memories and emotions are recruited into the system at the time, corresponding to different networks of the brain's neurocircuitry. This spatiotemporal logic can further extend to the rest of the brain, peripheral nervous system, and even to the entire world you observe at the time. Further evidence of this can be found in patients with dissociative identity disorders (DID), who have distinct or split identities—two or more selves, as in the famous case of Sybil. Thus, the same brain can have multiple regions that each experience a different "me." In such cases, a large portion of the neurocircuitry associated with each entangled system may overlap, and the distinctiveness—i.e., the different "me"—may arise because different memories and areas of emotion are recruited at different times. Sybil might be "Peggy" now, "Vicki" tonight, and "Sybil Ann" tomorrow, depending upon the areas of the brain that are entangled at any given moment.

We can actually observe the process, because analogous experiments have been performed that nicely illustrate superpositions.

In a 2007 experiment published in the journal *Science*, scientists shot photons into an apparatus and showed that they could retroactively alter whether these photons behaved as particles or as waves. The photons had to "decide" what to do when they passed a fork in the apparatus. Later on, after traveling nearly fifty meters past the fork, the experimenter could flip a switch . . . and whether or not they did determined how the particle had behaved at the fork in the past.

This type of "delayed-choice" experiment was first proposed, decades before it could actually be performed, by eminent Princeton physicist John Wheeler (Einstein's colleague, who also gave us the terms "black hole" and "wormhole"). You can see how it works in the following figure. If you follow the photons' path from the lower left, they first encounter a beam splitter. This beam splitter is the "fork"—if acting as particles, half the photons in the stream of light will proceed straight ahead, while the other half are deflected upward. A photon acting as a wave, on the other hand, would travel both paths, as discussed earlier in the book. After this beam splitter, there is an equal probability of each photon reaching one or the other of the detectors at the end of the experiment. If many bits of light are shot into the apparatus, when acting as particles, half will end up at one detector and half at the other. However, a second beam splitter—the dotted line at upper right—lets the paths be recombined into a single beam of light displaying interference effects characteristic of light's wave nature. Whether the experimenter chooses to turn on this second beam splitter determines how the photons exit the apparatus—in other words, it retroactively determines the which-way-path decision and particle-versus-wave decision the photon previously made, showing that events that already occurred can be altered by actions and observations made in the future. However, according to Wheeler himself, the "retroactive" interpretation of the delayed-choice experiment is somewhat misleading. Instead, he maintains that the experiment simply shows that the logic of what occurs at the fork (i.e., what happened in the apparatus in the past) depends on whether the second beam splitter is on or off—in other words, that nothing is collapsed until the second choice/observation is made in the present.

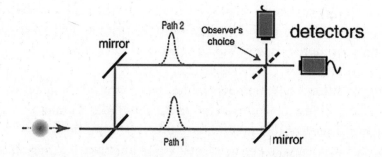

Fig. 7.1 Experimental realization of Wheeler's delayed-choice experiment. In 2007, scientists shot photons into an apparatus (arrow, lower left) and showed they could retroactively alter whether the photons behaved as particles or waves. The particles had to "decide" whether to take "Path 1" or "Path 2" at a fork in the apparatus. Later on (nearly fifty meters past the fork), the experimenter could flip a switch and turn a second beam splitter on or off ("Observer's choice," upper right). It turns out what the observer does at that point determines the logic of how the particle behaved at the fork in the past.

Whichever way you interpret it, the 2007 experiment and others like it seriously call into question whether there is a "fixed past." Indeed, since the 1960s, theoretical physicists like Wheeler have expressed the firm conviction that the past does not arise until the relevant objects are being observed in the present (more about this in Chapter 12).

Similar quantum effects in the brain strongly suggest that decisions, and even the mere fact of awareness, cause an entire cascade of quantum consequences that can even seemingly "overwrite" previous configurations. The important point here is that what's in your consciousness *now* collapses the spatiotemporal logic of what happened in the past.

Before we wrap up our discussion of the mechanisms of consciousness, one final can of worms deserves mention—namely the problem that arises as we attempt to describe someone's consciousness by reference to the activity of neurocircuitries in her brain.

If a scientist inspects the brain activity of somebody else, say Alice, then Alice's brain and its functioning is represented in the scientist's brain and awareness. So this attempt to probe the outside world, which includes Alice's mental functioning, is all still firmly planted within the scientist's consciousness. True, one can gain significant insight into how Alice's consciousness (or more precisely, our perception of her consciousness) operates in association with such activity. Yet however hard a scientist tries to understand Alice's consciousness and her perception of the outer world, the result is still just a picture or representation of Alice's brain.

In our attempts to understand another person's or animal's consciousness, we may also try to mentally put ourselves "in their place." Yet my feelings and thoughts remain singularly fixed to the one familiar consciousness I have always known as "myself." We never experience multiple consciousnesses, ours and that of somebody else. However comprehensive the information we have, when it comes to another's consciousness, we are at best seeing a picture within a picture, a play within a play—a mind that is only represented within our own.

Thus life offers different, hierarchical levels of representation. On the highest level there is a representation or "picture" of the world as perceived by consciousness, which can be either thought of as a delocalized state (the experience of absolute oneness) or the state that I experience as being centered in my brain. Within this highest-level picture are lower-level pictures or representations associated with other observers. This is illustrated in figure 7.2.

Fig. 7.2. Alice's representation ("picture") of the world is just a representation within Bob's representation of the world.

The "hard problem of consciousness" arises when we do not take into account and distinguish between these different levels of representation. Within the standard materialistic paradigm in which matter is primary, the hard problem is our inability to understand how experience, or perception, or feeling, can arise from insentient material objects such as molecules and brain tissue, or even the electrical pulses within them. However, within an alternative biocentric paradigm in which consciousness is fundamental—an axiom in which the "outside" world (and thus matter) is a representation in consciousness—the problem of how to derive consciousness from matter does not exist. We struggle to understand how consciousness arises from the brain of a person under our scientific investigation, but any awareness or representation of the world that we investigate is already within consciousness.

One can never fully explain consciousness as the "first-person experience" that we all recognize as that most familiar and intimate sense of "me." My consciousness (my first-person experience) is at a different level than an image of another person's consciousness that I can observe by studying the neural processes of his or her brain. Their consciousness, to me, is like a picture within a picture, as shown in the previous figure. It is not the true "me" experience. Therefore, all such studies stand apart from the true enigmatic "me" feeling. After all, a cat from a picture cannot eat a mouse in the room.

Fig. 7.3 An illustration of two different levels of representation: a room with a mouse on the armchair and a cat in the picture. That scene, together with the screen on the desk, is filmed by a boy, and the signal from the camera is transmitted into the computer, which then displays the image on the screen. The result of such self-referential loop is an infinite repetition of the picture within the picture. An analogous situation would occur if you were observing the functioning of your own brain.

Or can it? In the fascinating book *Gödel, Escher, Bach: An Eternal Golden Braid* by Douglas Hofstadter, there is much discussion about the tangled hierarchy of representations, with the example of a famous painting by Escher, *Print Gallery*, in which an observer regards a picture of a town that contains the gallery, which contains the observer himself.

If am observing my *own* neural processes, and, for instance, watch them on a computer screen, then I am involved in a self-referential loop, in which I see how my own consciousness mutates according to those neural processes. I am thus experiencing my consciousness experiencing my consciousness. It's rather like a serpent eating its own tail, which is a symbol from ancient Egyptian iconography. In the first known Western version, the serpent encloses the Greek words *hen to pan*, (ἓν τὸ πᾶν), meaning "the all is one." Its black-and-white halves presumably represent the Gnostic duality of existence.

But let's exit this hall of mirrors for the time being. We'll leave aside Egyptian iconography and the wisdom of the ancient Greeks in concluding "the all is one" to summarize what we've discovered so far by reviewing the seven founding principles of biocentrism.

And now we will add a new one, an eighth—the first of four additional principles this book will unveil.

PRINCIPLES OF BIOCENTRISM

First principle of biocentrism: What we perceive as reality is a process that involves our consciousness. An external reality, if it existed, would by definition have to exist in the framework of space and time. But space and time are not independent realities but rather tools of the human and animal mind.

Second principle of biocentrism: Our external and internal perceptions are inextricably intertwined. They are different sides of the same coin and cannot be divorced from one another.

Third principle of biocentrism: The behavior of subatomic particles—indeed all particles and objects—is inextricably linked to the presence of an observer. Absent a conscious observer, they at best exist in an undetermined state of probability waves.

Fourth principle of biocentrism: Without consciousness, "matter" dwells in an undetermined state of probability. Any universe that could have preceded consciousness only existed in a probability state.

Fifth principle of biocentrism: The structure of the universe is explainable only through biocentrism because the universe is fine-tuned for life—which makes perfect sense as life creates the universe, not the other way around. The "universe" is simply the complete spatiotemporal logic of the self.

Sixth principle of biocentrism: Time does not have a real existence outside of animal sense perception. It is the process by which we perceive changes in the universe.

Seventh principle of biocentrism: Space, like time, is not an object or a thing. Space is another form of our animal understanding and does not have an independent reality. We carry space and time around with us like turtles with shells. Thus, there is no absolute self-existing matrix in which physical events occur independent of life.

Eighth principle of biocentrism: Biocentrism offers the only explanation of how the mind is unified with matter and the world by showing how modulation of ion dynamics in the brain at the quantum level allows all parts of the information system that we associate with consciousness to be simultaneously interconnected.

LIBET'S EXPERIMENT REVISITED

*I am no bird; and no net ensnares me; I am a free
human being with an independent will.*
—Charlotte Brontë, *Jane Eyre*

We will now dive into one of the most ancient and fundamental questions of human existence—whether we have free will. Most readers may well consider this a waste of time, because . . . of course we each have free will! Didn't you just decide to order tuna on rye instead of the mozzarella-and-tomato salad? But let's take a deeper look. Remember, since the time of Descartes, scientists have largely considered the world to be controlled not by the whims of the gods but by physical laws and forces—things like inertia and gravity, and later, on the subatomic level, the rules of quantum theory. No matter what you believed about how the cosmos came into being, it was regarded as operating now like a giant machine following laws of cause and effect. These laws operate within our bodies, too.

So, if you cannot personally control the electrical firings within the neurons of your own brain, in what sense did you "decide" to get the tuna? When you really think about it, whatever pros and cons you might have considered, didn't the final decision at some level simply pop into your mind? At the very least, you have experienced yourself making other decisions that feel this way. And if you don't truly know *how* you made a decision, or *why* it happened, how can you claim to have exercised free will?

Well, okay, but if we start believing that things mostly happen on their own, how can we hold criminals responsible for their actions? Or motivate anyone to accomplish great things? What happens to our ideas about morality—and humanity in general?

This is obviously a much deeper and more complex issue than it might first have seemed. Even Einstein lost a lot of sleep over it. He was fond of quoting the nineteenth-century philosopher Arthur Schopenhauer, who himself liked to say, "A man can do as he wills, but he cannot will as he wills."

The fact that we're bringing this whole mess up—seemingly out of the blue—might well lead you to suspect that quantum mechanics or biocentrism will enter into it to clarify things in some major way. And you'd be right. Specifically, they will come to our aid when we now consider the famous Libet experiments, which are traditionally interpreted as a proof that we have no free will. This conclusion was based on the result of his cleverly designed laboratory setup, in which detection of the electric signal from brain activity repeatedly indicated that test subjects' decisions were made before they were even aware of their choices!

Nearly forty years ago, Dr. Benjamin Libet set out to discover whether the brain's autonomous electrical circuitry runs our lives "on its own" while meanwhile informing us of its decisions, which we usually then feel and assume to have been made by our sense of "me." Or, instead, whether the "me" sense truly steers our ship, as most of us have always assumed. Libet knew his results could have profound implications and might even settle ancient debates over individual free will once and for all.

Libet's first experiment, in 1983, consisted of three key components: a choice to be made, a measure of brain activity during that decision process, and a clock.

The choice, subjects were told, was to move either one's left or right arm, either by flicking one's wrist or raising a left or right finger. Subjects were instructed to "let the urge [to move] appear on its own at any time without any pre-planning or concentration on when to act. The precise time at which you move is recorded from the muscles of your arm."

The second component, the measure of brain activity, was obtained via electrodes on the scalp. Separately detecting the urge and the actual motion on the right or left was, fortunately, well within the experiment's abilities, because when electrodes are placed along the middle of the head over the motor cortex, characteristic electrical signals appear as one plans and executes a movement on either side of the body.

The clock was specially designed to let participants pinpoint sub-second times, and subjects were told to use the clock to report exactly when they made the decision to move.

Physiologists had known for decades that a fraction of a second before you actually move, there is a change in the brain's electrical signals. So, unsurprisingly, in Libet's experiment, the electrodes reliably recorded a change in brain activity a fraction of a second before participants moved. So far, so good.

The explosive result came from what the researchers found when they looked at when participants reported the decision to move. Libet's team discovered that this "decision" always fell in the interval *between* the electric change in the brain (technically termed *the readiness potential*) and the actual movement.

They found that the "feeling" of deciding simply couldn't be a report of whatever was actually causing the motion decision. The electrodes typically recorded a change in brain signals up to three-tenths of a second before the subjective experience of making a decision occurred. And the signals detected by the electrodes were indeed accurate, since by studying them, experimenters could always predict which arm, wrist, or hand would eventually be raised—before the subjects themselves knew!

These results seemed to show clearly that decisions are made in the brain's neurocircuitry before you're even conscious of them—thus, no free will. In short, the brain decides something, and soon afterward you

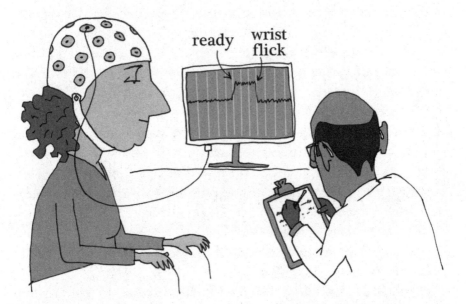

Fig. 8.1 The famous Benjamin Libet experiment is traditionally interpreted as a proof that we have no free will. This conclusion was based on the timing of an electric signal of brain activity that indicated a decision being made before the subject was aware of their choice. However, as we'll see, biocentrism arrives at an interpretation of the experiment that is opposite to the traditional, generally accepted one.

become aware of a decision, which you then (mistakenly) attribute to your own will.

This and later supporting experiments caused a big stir, provoking three front-page articles in the *New York Times* over the ensuing years and bringing the issue to a wide general audience. The *Times* pieces ended up concluding that there probably is no free will, but that society must pretend that there is in order to preserve the rule of law, hold people responsible for their actions, and so on.

In some circles, Libet's experiments were viewed with more of a shrug: If one part of the brain or mind makes a decision, even if the ego-circuitry that gives us our sense of being Nancy or George is merely passively informed about it, doesn't this still constitute a form of self-governance? After all, it's still our own brain running the show. Nonetheless, to most people, for whom the only self is the sense of "me," Libet's findings were humbling, if not outright upsetting. It appeared that our presumed status as captains of the ship of our lives was an illusion: Our kidneys cleanse the blood, our liver performs its five hundred functions, and the brain effortlessly makes all decisions on its own, including such everyday judgments as what restaurant to patronize and what to order when we get there. There was suddenly no place for Nancy or George, our sense of ourselves as conscious controllers.

But stop the presses and put away the antidepressants. There is good news for those unwilling to say farewell to conscious control: biocentrism supplies a powerful escape clause.

Quantum theory's many-worlds interpretation, unified with wave function collapse linked to consciousness—explanations that form the bones of biocentrism—gives us an alternate interpretation of the results of Libet's experiments. Namely, one in which we are *not* puppets whose actions are determined by proteins and atoms, but rather the active agent. From this perspective, it is solely my *conscious choice* that collapses the

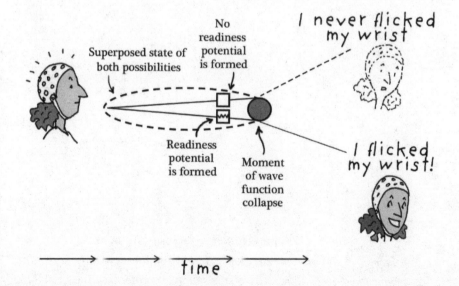

Fig. 8.2 Collapse of the wave function as perceived by a Libet-type experimental subject. After flicking a wrist, she finds herself in World 1. World 2 disappears from her perception.

wave function, and it does so at the moment I am aware of the decision to move my right or left hand. In other words, the collapse of the wave function does not happen at the time of the readiness potential detected by the electrodes. At that time, there is still a superposition of the possibilities, illustrated in figure 8.2 as different paths.

The interpretation of these experiments as demonstrating the lack of free will is based on the assumption that there is no distinction between the perspectives of the experimenter and the subject. The time order of events as seen by an external observer (the experimenter) of course indicates that the test subject, as perceived by the experimenter, made no choice: The decision was seemingly completed at the moment of the occurrence of the readiness potential as detected by the electrode. But this readiness potential was part of just one of the possible branches, and

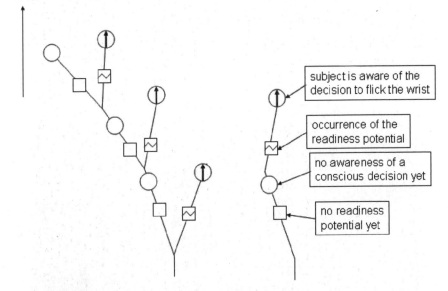

Fig. 8.3 A branching of the wave function relative to a third observer, who has not looked at the instruments (left). Relative to the experimenter (right), after having looked at the records of the readiness potential, the branching wave function collapsed into a single branch, in which there was first the readiness potential, and then the wrist flick (illustrated in the figures by the upward-pointing arrow).

from the perspective of the test subject, the wave function collapsed into that particular branch only at the moment she was aware of making the decision. All other branches then vanished from her perception.

The situation is different from the perspective of the experimenter. According to him, the wave function collapsed at the moment he viewed the result of the experiment. Before looking, there were several possibilities, while after looking, the course of subsequent events in the experiment is determined by the occurrence (or nonoccurrence) of a specific readiness potential.

From the perspective of each observer, the path their awareness will take is not predetermined. The same principle holds if a third person (à la "Wigner's friend" from Chapter 6) is involved. Relative to him, the entire setup—comprising the test subject, the experimenter, and the readings on the screen—is in superposition until he sees the result (figures 8.3 and 8.4).

Whether my awareness taking this or that path constitutes a conscious free will choice is a matter of definition. From my point of view, if I am a test subject in this experiment, my decision to flick a wrist or raise a finger is a free decision. Deciding in this moment to flick my left wrist means that the wave function of the world (including my brain)

Observer   Experimenter

Fig. 8.4 Illustration of how Libet's experiment is seen by a third person (looking through the window). For him, the entire setup—including the subject, the experimenter, and the reading on the screen—is in superposition until he sees the result.

collapses right now into the state in which, a fraction of a second before that moment, the matching readiness potential occurred. Had I decided not to move at all, my wave function would have remained in the state of both possibilities concerning the readiness potential (figure 8.2).

The traditional interpretation of Libet's experiment as implying that there is no free will falls within the paradigm of determinism. This is the paradigm mentioned at the beginning of this chapter and still defended by many scientists, one in which the universe is a great machine set in motion at the beginning of time, whose wheels and cogs turn according to laws independent of us. "Everything is determined," said Einstein, "the beginning as well as the end, by forces over which we have no control. It is determined for the insect as well as for the star. Human beings, vegetables, or cosmic dust, we all dance to a mysterious tune, intoned in the distance by an invisible piper." In this interpretation of Libet's result, every human thought, feeling, and action is the automatic and mechanical resultant of preexistent forces; the brain is a deterministic machine, its by-product being consciousness.

Even among those acknowledging the indeterministic realities of quantum mechanics, there are many who object that such indeterminism is for all practical purposes confined to microscopic phenomena, and others who would argue that quantum indeterminacy simply allows for actions to be a result of quantum randomness, which in itself would mean that traditional free will is absent, since such actions can't be controlled by conscious independent choice.

In the previous chapter we discussed the idea that quantum superposition extends to the workings of the brain, referencing in particular the theories of Henry Stapp. Stapp, among others, also argued that the quantum indeterminacy of brain processes enables an interpretation of Libet's experiment compatible with free will. In contrast to ours, Stapp's explanation does not rely on the many-worlds theory—instead it involves a detailed mechanism of brain processes leading to the readiness potential and to the subsequent conscious decision to move a finger. Stapp nicely explains why the brain cannot be a deterministic machine and that its processes are in quantum superposition.

Whether or not the brain can indeed be in a superposition state is a matter of debate within the scientific community. However, Stapp pointed out that quantum coherence in the brain is one thing, but the brain and the environment together are in a pure (that is, superposed) quantum state. And, in fact, the "environment" extends to the entire universe.

Hence, even if it turns out that quantum superposition does not occur on the level of the processes in the brain, this pure state means the quantum system that comprises the brain and its environment embraces many possible experiences of the observer, which are then "actualized" into one definite experience by the collapse of the wave function. The decision made by the test subject about which of the available experiences (and thus branches) to enter cannot unfold at the occurrence of the read-iness potential, since the subject is not even aware it is happening—at that time the wave function is still in a superposition state. The decision actually happens a bit later, when the wave function collapses.

To sum up, in the biocentric interpretation of Libet's experiment we've outlined in this chapter, *you* are the agent that collapses events. You determine the path you take within the branching tree of many possible paths, as illustrated in figure 4.4 of Chapter 4 and the figures of this chapter.

And it's not your subconscious, either, contrary to Libet's implica-tions. Your *sub*conscious is, exactly as the word implies, below conscious-ness, or as Wikipedia defines it, "mental processing that occurs below awareness, such as the pushing up of unconscious content into con-sciousness, and to associations and content that reside below conscious awareness but are capable of becoming conscious again." Obviously, the body does many things subconsciously, and it performs numerous invol-untary, reflex, and automatic actions in response to stimuli and without thought—such as yanking one's hand from a sizzling pan. But this does not mean that *all* our behavior is a result of subconscious activity and that consciousness has nothing to say. In the biocentric interpretation of Libet's experiment, your conscious awareness is what chooses just one of the available paths—which then becomes the reality you experience.

So there goes your excuse for coming home so late that the pot roast has long gone cold. "I couldn't help but stop for a drink," will no longer fly. You may have previously tried to blame it all on your readiness potential, having no free will of your own, but now that the wife has read this chapter, she is wise to you. Hands on hips, she's got you nailed:

". . . And next, I suppose, you'll be blaming it all on a wave function that collapsed while you weren't paying attention, right? Well, forget it!"

ANIMAL CONSCIOUSNESS

> *We patronize [the animals] for their incompleteness,*
> *for their tragic fate of having taken form so far below*
> *ourselves. And therein we err, and greatly err.*
> —**Henry Beston**

It's natural to be people-centric when we explore consciousness. We're all drawn toward the familiar. And as we've seen, we're just beginning to understand our own human consciousness, making it surely even harder to probe that of, say, an octopus. But subjective experience and the exquisite and varied processes that facilitate perception are indeed also enjoyed by creatures very different from us. They may possess neural architecture quite distinct from the structures of the human brain, architecture that is nonetheless clearly designed to enable consciousness to be centered or localized within it. The neural structures of an organism's consciousness are evolved to impart singular experiences tailored for specific situations and habitats.

As to how conscious experience manifests in nonhuman life forms, one broad difference might be familiar thanks to a recent development in primary education—an emphasis on "mindfulness," a practice dating back to ancient meditative traditions that (based on data showing its effectiveness at improving concentration) some teachers have now been trained to suggest to students. Those just now hearing the term may take it to suggest time spent in thought, but the practice of mindfulness actually involves the opposite. The idea is to be attentive to immediate sensory experiences rather than ruminating about this or that. If students can simply observe whatever they see or hear, paying attention to the unending details unfolding in the present moment rather than daydreaming, they will be sharper, keener, and derive more benefit from the here and now, including the classroom experience. Bottom line: the huge brains we've been given can be as much a distraction as a gift. This kind of being "in the moment" is the type of consciousness that, as far as we know, most corresponds to that of other conscious organisms.

By "other conscious organisms" we mean animals—including birds and insects—that have brains, sense organs, and appendages that allow them to locomote and move around in space, as well as animals and plants that do not actively move, but that can store memories and respond to their spatial environment.

Mindfulness might bring us more into sync with the experiences enjoyed by nonhuman animals, but the differences in our conscious experiences obviously run far deeper than the unique human penchant for daydreaming. Some organisms utilize sensory inputs that are entirely absent from our own awareness, or if present, have through time slowly degraded until they now play negligible roles in daily life. Once we begin to take a look at animal consciousness, we find ourselves in an almost endless exploration of strange new worlds. Remember, reality exists relative to a particular observer—animal consciousness, like human consciousness, involves the collapse of wave function. And the unique physiological setups of other animals allow their choices and wave function collapses to unfold along pathways that diverge from ours in wonderfully creative and useful ways.

Everyone who has ever had a dog knows where most of a canine's attention is centered. On smells, of course. And there's no need to speculate about whether this proclivity is mere habit or has some deeper genetic and environmental dictates. Just look at Rover's face. Check out that nose! It starts just below the eyes like ours does, but then extends halfway to Florida. Is it any surprise that 90 percent of Rover's attention is on environmental chemistry?

To smell something, at least one molecule of the substance must land on and cling to the moist mucus membrane that lines the nose. (This is why some very large molecules like tetracycline and DNA have no smell at all: they're just too big to stick to our noses.) Dogs with very sensitive olfactory abilities can detect just a few molecules wafting in the air. Researchers estimate that a bloodhound's nose contains 230 million olfactory cells, which is forty times more than we humans have. And while our brain's olfactory center is the size of a postage stamp, a dog's can be as large as an envelope.

All of this sensory architecture doesn't merely give a dog the ability to discern barely-there odors; it lets the dog luxuriate in them. Their world is a melange of fascinating biochemical excretions that convey the rich story lines of creatures that have recently been in the vicinity. Why, then, should they share our own focus on the visual? Indeed, humans perceive a wider range of colors than canines—in the green part of the spectrum, where we are most sensitive, we can distinguish between fifty different shades of that hue alone. By contrast, dogs cannot detect *any* difference between green, red, and yellow—it's all a single shade to them, whose sole clearly contrasting tint is the color blue.

Many animals possess consciousness that creates vastly different visual experiences than those of humans, whose visual acuity is better than most. This is illustrated in the following figure. We humans see the White House as depicted in the upper panel (though usually in color!), whereas certain insects would collectively collapse a reality closer to that seen in the lower panel.

With such a lack of visual variety, why should Rover want to stare when he could sniff? But our dog's consciousness diverges from ours in

Fig. 9.1 The top panel shows how humans, with their acute visual sense, see the White House; the bottom panel shows how it might appear to insects.

ways more dramatic than time spent on sights versus smells. Dogs have recently been shown to sense magnetic fields!

We've long known that some animals navigate by aligning themselves with Earth's weak magnetosphere, a barely-there force of just 0.5 gauss. Bees, birds, termites, ants, hens, mollusks, many bacteria, homing pigeons, chinook salmon, European eels, salamanders, toads, turtles—the list is long. These creatures have magnetotactic abilities, in some cases caused by their central nervous systems responding to chains of magnetosomes, which are tiny specks of iron-rich minerals like magnetite surrounded by membranes of fatty acids and, typically, more than twenty proteins. This architecture is so wondrous, and produces a sensitivity so acute, that some animals create a mental plot of subtle variations in our planet's magnetic field, providing an internal road map of their location. In other cases, magnetism serves as a backup navigational system, as it does for some birds when the sky is overcast so that the sun and stars are hidden.

The fact that dogs might also exhibit this kind of magnetic talent was suspected long before it was proven because of their curious preference for relieving themselves with their bodies aligned north-south. What's more, their fellow canines, red foxes, had been observed for centuries to exhibit an odd directional preference of their own when pouncing on prey. If you've ever seen a fox making its distinctive high leap upon a vole or mouse or some apparently empty snow-covered spot where they've detected a sound emanating from the subnivean realm (the often vacant space between the ground and snow cover), you were probably not paying attention to the cardinal points of the compass . . . but if you had been, you'd likely have seen the fox leaping toward the northeast.

No matter what biome or environment a creature inhabits, nature's innovativeness in meeting its challenges and conferring advantages seems virtually limitless. Take, for example, the detection of infrared, or heat.

We humans have skin that can sense when a nearby object is hot. But this ability only operates when the object is hotter than 109.4°F, or 43°C. In contrast, vampire bats can detect heat at distances up to eight inches and at temperatures as low as 86°F—a range that covers the skin warmth of virtually every mammal on which they might wish to perform their Dracula routine.

Bats, of course, are most famous for a different sensory ability, one that's even more alien to us. This is their sonar mechanism, in which they chirp a continuous series of sounds and then detect sonic reflections that reveal the distance to some flying prey or a cave wall they wish to avoid. They can even get information about a target's movement by discerning the echo's Doppler shifts, similar to the pitch changes we perceive when a car horn or ambulance siren is moving toward or away from us. These sophisticated talents are impressive enough, but echolocation abilities reach astonishing perfection in toothed whales and dolphins, whose sound pulses can penetrate soft tissue to provide them with an X-ray-like mental image of the object of interest.

Dolphins have still more up their little sleeves. They have the ability to *reproduce* the echoes of their own sonar signals, so that when they have found something interesting, like a delicious school of juicy fish, they can replicate the sounds to "tell" other dolphins what they've discovered. In doing this, they don't employ the kind of clumsy, symbolic, one-word-at-a-time process we humans use for communication. Instead, they actually create a visual picture of what they just saw in the minds of other dolphins, perhaps even "bolding" or "highlighting" aspects they wish to emphasize.

Yet another perception technique we poor humans lack is the ability to perceive electrical fields. Much has been made of the putative health risks for people whose homes are adjacent to high-voltage power lines, which are enveloped in huge electrical and magnetic fields. The electrical fields around power lines—and even around our appliances and computers at home—are produced whether or not a power-using device is turned on, whereas magnetic fields are created only when current is flowing. Major power lines produce magnetic fields continuously because current is always flowing through them. Electric fields are easily shielded or weakened by many objects, such as intervening walls, whereas magnetic fields can pass through buildings and most other materials, as well as living things. Many have speculated about how the human body might be affected by being bathed in such a strong field nearly 24/7. Although studies have been somewhat inconsistent, it seems that those exposed to the strongest fields (above 3 or 4 microteslas) suffer a small increase in the risk of some cancers. Whatever the specifics, our animal bodies are known to be affected by electromagnetic fields, which do not simply pass harmlessly through us like neutrinos. And so it makes sense that, for creatures with physiological architecture designed for the purpose by evolutionary processes, such fields might be consciously detectable.

In other words, we shouldn't be surprised to learn that sharks have organs called Lorenzini blisters, which sense electrical fields. This electro-reception ability is shared by several sea creatures, but only one mammal, the platypus. Bees can sense electrical fields as well, though they do it in a roundabout way: they accumulate a positive electrical charge during flight, and then the negative charge often present in flowers makes the hairs on the bee's legs stand on end, alerting them to the presence of flora. (Bees get further help from eyes that, unlike ours, can see ultraviolet wavelengths. It turns out that many flowers flaunt gorgeous, intricate patterns that are only visible in ultraviolet light.)

So far, we've mostly explored the ways animal consciousness can operate by detecting what to us are invisible emanations. But what about mechanisms for detecting more straightforward, tangible stimuli—like actual substances hitting us? One such mechanism gives rise to what we experience as sound.

The underlying nature of acoustic experience continues to be misun-derstood by most people. Evidence of this was seen when, after a lecture to a general audience, one of the authors asked what is probably the world's oldest and most basic question involving consciousness:

"If a tree falls in a forest, and no person or animal is present to hear it, does it make a sound?"

The audience was asked to vote yes or no by a simple show of hands. The result? Some three-quarters of the lecture hall voted yes: by consensus opinion, the tree does make a sound even if no sentient being is nearby.

This is the wrong answer. But it nicely illustrates the public's wide-spread confusion about sound—and indeed about consciousness in general.

When a tree falls, the physical fact of the massive trunk and countless limbs striking the ground produces disturbances in the air that envelops the scene. Rapid, complex pulsations in air pressure radiate in all direc-tions, diminishing with distance. In events involving weighty enough objects (like falling trees) or great enough force (like an explosion), these air pressure changes can actually be felt on the skin as quick puffs of wind, which is why deaf people can have no small sensual experience if seated in front of the main stage speakers at a rock concert.

These air puffs are the physical occurrence that results from a tree falling. In and of themselves, they are silent.

But when they encounter the eardrums or tympanic membranes of humans or animals, they physically impart motion to this thin layer of tissue. Attached neurons respond to the resulting vibrations in these membranes by sending electrical signals to the brain, where many billions of cells are triggered to produce what we humans or animals experience as specific sounds.

Thus, *the sound is coming from inside the house.* Noises are produced by our own neurons, which manifest their conscious experience. The noise of a tree falling is the end result of air pressure variations that push on tympanic membranes designed to wiggle in response, but obviously none of this—save the (itself silent) air disturbance—happens if nobody is in the woods that day. This isn't a philosophy lesson, but a straightforward fact of physics and nature: the falling tree, in and of itself, cannot make a sound, because a sound is by definition a conscious experience.

What each conscious organism does with a given set of vibration-producing wind puffs is another matter. Humans are sensitive to sounds at frequencies from 20 to 20,000 Hz; the perception of a sound by organisms sensitive to a wider or different range may be quite unlike our own. There is no way to know whether what we experience as the deep, low rumble of distant thunder is perceived by a cat as a high-pitched whine. The inarguability of the subjective nature of conscious experience is further proof that it is a symbiotic phenomenon, an amalgam of "external" nature and ourselves. Of course, to be strictly accurate, even the "external" world of stimuli has no definite and independent existence outside of consciousness. People and animals likewise have no existence independent of a conscious observer, even if they may be that observer themselves.

But let's return to sound. While there is much we cannot know about the subjective experience of sound in other animals, with observation and the help of technology we are slowly learning about how other organisms use sound. Many produce sounds deliberately for communication, as we do. Researchers have found that social insects like bees and ants typically use between ten and twenty separate recognizable vocalizations, while the number is three to four times that in social vertebrates like wolves and primates. And just as sound perception is variable, so are its methods of production. While many organisms communicate with vocalizations, others, like crickets, produce their sound-based communications via other means, such as rubbing their wings together.

A century ago, Tufts professor Amos Dolbear created a stir when—seemingly out of the blue, since it wasn't his field—he published an article in the *American Naturalist* that revealed that anyone can tell the temperature simply by counting cricket chirps. Quickly termed Dolbear's law, it became all the rage in naturalist circles and among campers. Though its details are definitely a bit off topic, you may as well be the only one on your block who possesses this singular temperature-finding skill. Ready?

Simply count the number of cricket chirps in 14 seconds and add 40. That's the current temperature in degrees Fahrenheit.* What could be simpler? And Dolbear's law is accurate to a single degree.

As for settling that old barroom argument over which organism has the best hearing (What? Maybe you don't go to the right bars . . .), the answer is the moth. Moths can detect even higher-pitched sounds than bats can—which is saying something, since the latter is the creature they're most desperately trying to evade. The bat comes in at number two, and after that the keenest hearing belongs to the owl, then the elephant and the dog, followed by cats. The worst hearing? Probably snakes, whose consciousness is naturally and understandably more attuned to ground vibrations than fluctuations in air pressure.

By now the point has been made: organisms' awarenesses are fine-tuned to be sensitive in a myriad of ways and by utilizing a range of physiological structures. Each is left with theoretical freedom to attend to reality through a wide variety of experiences, but, in truth, has had those freedoms focused and filtered by the complementary forces of environment and evolution so that, in practice, a much narrower selection of inputs is likely to occupy an organism's attention at any given moment.

Through it all, it's good to remember that though the animal body is the instrument of sense perception—like a big neuron antenna—all

---

* If you prefer Celsius, count the chirps in 8 seconds, and add 5.

sensory data is ultimately processed in the brain. The brain receives nothing but impulses, a ditdit ditditditdit ditditdit of electrical signals carried from the senses to the nerves. The brain receives broken-down information and has to put this disjointed mass of data back together, which it does according to very specific laws. It reassembles sensory data according to the rules of time and space—the logic of the brain.

Time and space are projections created inside the mind, where perception, feeling, and experience begin. They are the tools of life, the representations of intellect and sense that even the smallest turtle hatchling must learn to use once its glistening eyes open for the first time. The hatchling, wandering solo on land through leaves of sweet fern and seed heads of bluestem grass, sometimes traveling for more than a week before settling into a pond or swamp, must rely upon these tools to navigate the world.

All animals with nervous systems have some of the same basic machinery. This is not a fluke. Spatial and temporal understanding certainly exist in other animals besides humans, although the "wattage" and "instrumentation" of our senses may differ. We might think of "wattage" as how bright or dim a sense is: a hawk has acute eyesight, able to process a tremendous amount of visual information; an African mole rat is blind, and like many cave-dwelling creatures, has no organs for registering light. Thus, the visual sense perception of a hawk "burns" at a high wattage, while a mole's is dim to dark.

Sight, smell, hearing, touch, and taste are our familiar human sense "instruments." Various animal species share various of these five senses in various wattage intensities, and as we've seen, may also employ other senses we might find hard to intuit. Most insects, for example, don't hear as humans do but instead feel vibrations—often through sense organs in their feet—as constant tremors. The vibration-sensitive "ears" of a field cricket are located in its knees. We've already seen how some bat species navigate their world not by sight or smell, but by echolocation. Schooling fish are highly sensitive to water pressure along a "lateral line" on each side of their belly, enabling them to synchronize their own movements with those of other fish in close proximity and thus move as a unified, fluid whole.

In biological terms, the logic expressed in the circuitry of the brain is linked to the logic of the peripheral nervous system. They are coordinated. The differences in wattage and instrumentation among animal species circumscribe the universe distinctly for each.

Animals and humans are able to discern multiple sense perceptions as existing alongside one another at the same time, observing them as objects existing outside us and as occurring in space. A human being, for example, might perceive the scent of lilacs bursting from bright spring clusters poking through a chain-link fence from a fertile backyard into an alley where trash cans overflowing with ripe garbage reek in the pale light of an overcast sky while a plane roars overhead. And yet for all these conscious sense-mediated experiences—a potpourri of unending sensations—we humans sometimes place ourselves in a radio-static mode, attuned to no sense whatsoever, lost in the internal world of our thoughts until we suddenly realize a friend has been speaking . . . and wonder if that "mindfulness" business might not be such a bad idea.

As far as we know, humans are the only animals who cease attending to their external awareness in this way, attending instead to our own thinking—or even, as you've done while reading this book, thinking *about* thinking. There is no doubt that the consciousness of animals differs from ours, perhaps in ways we can only guess at. And as a result, their realities—which are after all derived from their first-person experience as observers—differ as well. Yet there is a sense in which these differences are illusory. Consciousness and wave function are experienced as localized in our particular brain, creating the sense of "I," the so-called "me feeling," but as we discovered in Chapter 5, the lack of separation proved by entanglement experiments suggests that my consciousness and your consciousness, or your consciousness and that of your dog Rover, are in fact manifestations of a single consciousness.

One of the authors, Lanza, recalls contemplating the implications of this oneness:

I remember fishing on a warm summer night. Now and then I could feel the vibrations along the line linking me with the life prowling about the bottom. At length I pulled some bass, squeaking and gasping, into the air.

In experiments, it has been repeatedly shown that a single particle can be two things at once. Physicist Nicolas Gisin sent entangled photons zooming along optical fibers until they were seven miles apart, then measured one and found the other "knew" the result instantaneously, suggesting them to be intimately linked in a manner only possible if there is no space between them, and no time limiting the speed of their communication. Today no one doubts the connectedness between bits of light or matter, or even entire clusters of atoms. See the loon on the water, the dandelion in the field. How deceptive is the space that separates them and makes them appear solitary.

In the same way, a part of us is connected to the dandelion, the loon, the fish in the pond. It is the part that experiences consciousness, not our external embodiments but our inner being. According to biocentrism, our individual separateness is an illusion. Everything you experience is a whirl of information arising in your brain. Space and time are simply the mind's tools for putting it all together. However solid and real the walls of space and time have come to look, inseparability means there is a part of us that is no more human than it is animal. And as parts of such a whole, there is justice. The bird and the prey are one. Make no mistake about it: it will be you who looks out of the eyes of your victim. Or you can be the recipient of kindness—whichever you choose.

This was the world that confronted me on that warm summer evening. The fish and I, the predator and the victim, were one and the same. That night, I sensed the union every creature has with every other. In the words of an old Hindu poem: "Know in thyself and All one self-same soul; banish the dream that sunders part from whole." The consciousness behind the youth I once was and the man I became was also that behind the mind of every animal and person existing in space and time.

This may not unsettle you, except perhaps on a warm moonlit night with a fish gasping for life at the end of your rod.

"We are all one," wrote Loren Eiseley, the noted anthropologist, "all melted together."

I let the fish go. With a thrash of the tail, it disappeared into the pond.

QUANTUM SUICIDE AND THE IMPOSSIBILITY OF BEING DEAD

10

In the beginning there were only probabilities. The Universe could only come into existence if someone observed it. It does not matter that the observers turned up several billion years later. The Universe exists because we are aware of it.
—**Martin Rees**

W*hy am I here?* It is a question most everyone has asked at one time or another, often late at night or in the wee hours of the morning. It may not seem like the sort of thing science is most suited to shed light on, but in fact the question of why you happen to exist instead of *not* existing is intimately related to the physics we've been exploring throughout this book.

In the quest to decipher how the universe works on the most fundamental moment-by-moment level, one persistent stumbling block has long been explaining why one event happens rather than another. With the advent of quantum theory, it became clear that an experimenter had equal chances of observing an electron whose spin was "up" as opposed to "down." But determining why the experiment unfolded one way and not the other seemed impossible.

Niels Bohr, in the 1920s, offered what became known as the Copenhagen interpretation, which, as we've seen, essentially said that all possibilities hover invisibly over the experimenter and his lab in the form of a "wave function." The act of observing, said Bohr, causes this wave function to collapse, which means that the multiple possibilities suddenly vanish in favor of one definite result. But for all its revolutionary insight about how the uncertain quantum world becomes definite reality, this interpretation had no answer for the question of why, in a case where both have equal probability, one reality should emerge instead of another.

Then, in his 1957 doctoral thesis, Yale graduate student Hugh Everett proposed a remarkable alternative in which no particular single collapse need occur—because in fact *every* option occurs. He posited that instead of wave function collapse, the universe branches into separate forks so that all possibilities unfold. The observer is part of the fork or branch in which he observes the electron with an "up" spin, but a separate copy of himself sees a "down" spin and then continues his life with memory of that.

You will recognize this as the many-worlds interpretation (MWI), already discussed at some length in other chapters. But as biocentrism essentially offers an improvement on Everett's original interpretation, it's important to continue our exploration of this radical change in how we might view the cosmos. Not least because, as we'll see in this chapter, it is the key to untangling questions of life and death.

We'll start with the self-evident fact that consciousness is not a tentative, on-and-off kind of thing. Consciousness, according to biocentrism, is fundamental to the cosmos and impossible to separate from it. We see this firsthand with our own experience of cognition, in that it never disappears. Some might ask, "What about when you die?" But experiencing "being dead" is a logical paradox—you cannot simultaneously "be" and also "not be." One of the properties of consciousness is that it is never subjectively discontinuous. You cannot experience nothing, since even the words "experience" and "nothingness" are mutually exclusive.

A model of how this works in the context of the MWI is neatly illustrated by the so-called "quantum suicide" scenario, in which a gambler playing quantum Russian roulette always feels himself surviving.

Let us envisage this experiment, nicely explained by theorist Max Tegmark. A professor, who is a determined believer in the many-worlds interpretation of quantum mechanics, gives to his assistant a special quantum gun, and instructs her to fire successive shots at him. A given pull of the trigger will either instantly snuff out his existence or cause the gun to emit nothing but a loud "click." If, instead of firing, the gun only makes a "click," the assistant must shoot again, and so on until the gun actually discharges.

In this experiment, there are two perspectives. From the point of view of the assistant, after a few trials she is horrified to see that she has killed the professor. But from the point of view of the professor, the gun never fires. At every attempt there is only a click. This must be so, because unlike the actual game of Russian roulette, which uses an ordinary revolver with only one bullet, quantum Russian roulette uses a gun that operates on the principle of quantum superposition. Before every pull of the trigger, the gun is in a superposition state of "click" and "fire." Because the professor is intimately linked with all this, the initial state is composed of the gun in its superposition state and the professor in a definite state of being alive. After the first shot, this initial state evolves into another superposition state for these two components, one state with "click" and "professor alive," and the other with "fire" and "professor dead." Let us illustrate this in symbols:

$$(|click\rangle + |fire\rangle)|alive\rangle \longrightarrow |click\rangle|alive\rangle + |fire\rangle|dead\rangle$$

These two states—one in which the gun emits a click and the professor remains alive, and another in which the gun fires and the professor is dead—are each branches of the superposed wave function, constituting two Everett worlds. The professor's consciousness, by definition, cannot enter the world in which he is dead, and so at every shot it jumps into the branch/world in which his brain is intact—that is, in which the gun did not fire. Everett himself was intrigued by the idea of such an experiment. However, he did not perform it, noting that even if from his perspective he remained alive, there would be many worlds in which his relatives would be sad at learning of his death.

In a way, each of us plays a version of quantum roulette every day, at every moment of our lives. Namely, wave function contains many possible outcomes (Copenhagen view) or branches (many-worlds view). From our first-person's perspective, each time a choice of outcomes unfolds and wave function collapses to reveal a single result, we always find ourselves in an available world that supports consciousness. For we are perpetually aware of something, with no intrusive gaps in which oblivion makes an appearance. Even our memory track, when conjured to play its cherished recorded recollections, contains earlier and earlier—though usually less and less detailed—"home videos" from ever-younger periods of our life. At a certain point in the past we can no longer picture anything, but this does not mean there was nothingness at that time, only that very young brains lack the ability to retain distinct memories. So memories are not reliable markers of conscious experience; particularly relevant to consider may be times when others insisted we were unconscious for a period of time, having fainted or similar. But to us during such experiences, no time at all passed; we felt woozy and the next thing we knew we were "coming to." No experiential gap has ever occurred for us. And if this has been true even during deep comas, why do so many fear that death will bring the arrival of nothingness?

It has been pointed out in scientific literature that if MWI is valid, then from any individual's perspective, one always finds oneself alive so long as there is a branch/world available in which one's own body's structure supports consciousness. However, over the span of your subjectively perceived unfolding of life, the number of further branches or worlds supported by a brain configuration in which you are older decreases as you travel the course of a particular branch of your life. If you are 140 years old, for instance, then there can be no Everett world that would lead you to feeling yourself becoming even older. When no such "living" branch remains, then the wave function, together with the consciousness it is associated with, can no longer be localized/centered in your particular brain configuration—but nor can it cease to exist, since wave function, in common with all other fundamentals of nature, cannot vanish.

According to Everett's MWI interpretation, many *other* possible configurations are, and always will be, available to support your consciousness, including a world in which you find yourself two years old and living a slightly different life, that is, an alternative history.*

In quantum mechanics, a localized wave function, if not observed, spreads through the entire universe. In fact, according to the MWI interpretation, it spreads over the multiverse because it contains all possible positions of the particle, and each position belongs to a different Everett world. However, QT tells us that if immediately after a particle is observed you observe its position again, the particle remains localized at that position or at a nearby point. Thus, if the "wave packet" is permanently observed, it remains focused at one position. The same must happen with a "large wave packet"—or more specifically, one that corresponds to the macro world of your human consciousness. That wave function contains many degrees of freedom embracing numerous particles, atoms, molecules, proteins, organs, and so on—all coupled to "external" degrees of freedom such as those making up the environment. Such wave functions are entangled systems performing continuous self-measurements or observations.

However, when all this majestic structure associated with your current awareness is broken by an outcome in which there is no Everett world that allows your consciousness to continue functioning within that particular body/brain configuration, then measurements, observations, and self-reflections are no longer possible along your existing course, and the wave function spreads in a way that is analogous to an unobserved single particle wave packet. Then, just as a solitary wave packet collapses into a definite position upon reobservation, so our brain-associated quantum wave packet collapses into another world of definite experience. This

---

* For a fictional depiction of this, one might revisit the fascinating 1998 German movie *Run Lola Run*, described in this context in the book *Biocentrism*.

could be you at a different age, or in a different Everett world in which you made alternate decisions.

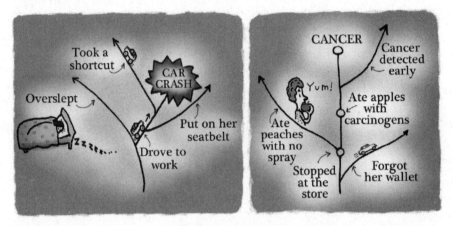

Fig. 10.1 Examples of possible personal histories. In one branch there is a tragic event (left: car accident, right: death from cancer), whereas in other branches, the person survives. At every cross point, consciousness hangs up on one of the branches in which life is possible. For instance, one of the authors' sisters died in an auto accident, but according to the MWI, that was not the end of her consciousness—it continues along one of the other branches.

The enigmatic issue of death should therefore be understood within the thesis that wave function, relative to an observer and representing his experiences of the world that he lives in, can never cease to exist, and that from an observer's first-person perspective, there is no death. The observer is always aware of something.

In the worldview adopted here, there is only one consciousness—it can be localized/centered in a particular brain configuration and thus experience the world from that particular point of view. Alternatively, it can be localized/centered in a different brain configuration and experience the world from that different perspective. Localization of consciousness in a particular brain is the result of observer-dependent wave function collapse. Just as consciousness finds itself in one of your Everett branches (but could have found itself in some other consciousness-supporting branch), so it finds itself in one particular brain (but could as well have found itself in some other brain and experience the world from that other point of view).

Consciousness, localized/collapsed in another person, experiences a different world than consciousness collapsed in me, because it has different thoughts, different experience of body motion, different details of the environment, and so on. The difference between the worlds experienced by you and another person may be thought of like those experienced by different Everett versions of yourself. The world experienced by the current me is in many respects practically the same as the world that would be experienced by a different Everett version of myself—the same Earth, sun, continents, towns, relationships, and so on. Depending upon similarities of environment and so on, the same could be said for the worlds of other observers. In other words, worlds associated with different observers are analogous to Everett worlds.

Now, if we adopt the view that Everett's alternate worlds are real (whatever this means), then it follows that the worlds in which wave function has collapsed into different brains (including those of animals) are also real. In such a way we avoid solipsism. Indeed, reality is created by the observer, but there are in fact many realities, each observer dependent. If we assume that our alternate Everett worlds are only possibilities, and the real world is solely that of our current experience, this implies that the branches of the universal wave function localized/centered in other brains are likewise not real, but merely possibilities. Therefore, denying the reality of Everett's many worlds means accepting solipsism, while accepting the reality of Everett's many worlds leads to refuting it.

One last point, an important one, presents itself here. It may seem as if we are saying that awareness can "jump" from brain to brain. But jumping in the usual context implies that time and space are absolute, external things. In truth, except for what you're experiencing now, everything else exists for you in superposition. "Time" or "space" can only be experienced relative to an individual observer. Independent of the observer's consciousness, space and time are nonexistent, which means there exists no linear connections outside of consciousness. All branches are superpositions within consciousness, and upon wave function collapse, consciousness finds itself in one of the branches.

Besides giving us a new way to look at the unfolding of our lives, the ideas related to wave function, many worlds, and consciousness that we've discussed in this chapter can also be used to view the evolution of the universe in general, and life on Earth in particular, from a unique angle—one that explains why you and I are here now despite the overwhelming odds against it. As we'll see, a quantum suicide-like argument does a much better job at this than the standard "Dumb Universe" model, which tries to argue that a cosmos as numb and insensate as shale came up with people and hummingbirds by randomness alone.

In addition to the two hundred or so physical parameters that must be exactly as they are if, on basic chemical and physical levels, life-friendly conditions are to be found around us, there's the whole business of life's creation in the first place, with its own lengthy rider of Goldilocks-like requirements. A planet that is neither too hot nor too cold, for instance. Or, you know, radiation filled. Even here on Earth, life would be close to impossible if we didn't possess our massive nearby moon—without it, our planet's axial tilt would wobble wildly, sometimes aiming straight at the sun to produce unlivably hot temperatures. Earth manages to avoid such chaos only because of our moon. And how did we get this moon? The perfectly timed collision of a Mars-sized body coming from a very specific direction and at exactly the right speed—not so fast or massive as to destroy us, and not so small as to fail to do the job. Direction matters because, as a result—unlike all the other major moons in the solar system—our moon doesn't orbit around its planet's equator. If it orbited "normally," it wouldn't exert its torque in the alignment needed to stabilize our axis. Another convenient accident.

The conventional, materialistic interpretation holds that our universe was born in the big bang and ticked away "out there" for billions of years until, by chance, on the planet we call Earth, life started to develop, and events happened to continue in a way that eventually led to the phenomenon of me being conscious of that universe. If this were indeed so, then the fact that I am alive and conscious (leaving aside the hard problem of how any consciousness arises from matter, discussed earlier) is the result of a long chain of extraordinarily well-tuned events.

Had that chain been slightly different at a single link, there would be no me and my consciousness. Not only must many things on the cosmological scale have happened exactly as they did, but once life emerged on Earth, it must also have evolved precisely as it did, and all my ancestors, not only human, but animal as well, must have survived all fights, diseases, accidents, natural disasters, fires, earthquakes, and so on; they must have been victors in all battles, survivors in all wars and at every occasion, managing to transfer their genes to their descendants until the chain gave birth to my body. If my parents had not met each other, I would not exist. Had they lived slightly differently, not me but a brother or sister might have been born. The typical male generates over half a trillion sperm during his lifetime, whereas the typical female generates hundreds of thousands of eggs. Only one of those trillions of combinations would have led to my birth. Yet, I had the unfathomable luck of winning this biological lottery. According to this view, I am here being conscious because this long chain of events unfolded just so; if it had unfolded otherwise, my body would not exist, and therefore neither would my consciousness. There would exist an external world with other people, but I would not be aware of it. And, if cosmological evolution had been just a bit different, there might be no habitable Earth at all, and perhaps no other habitable place in the universe. There would be a universe, but nobody would be aware of it.

We have seen that the universe does not work according to the conventional materialistic scenario sketched above. In fact, it is the other way around. Matter and a universe arise from a collapse of the grand wave function into a definite world of conscious experiences. Here is where the reasoning behind the quantum suicide experiment and the many-worlds interpretation offers us clarity: the universe supports your consciousness because it must. The chain of improbably fortunate events that leads to you experiencing the universe is one of the branches of the superposed wave function, collapsed relative to you, the observer. Just as the "quantum suicide" professor cannot find himself in a branch in which the gun fires, you cannot find yourself in a branch that lacks a chain of events ultimately supporting your consciousness.

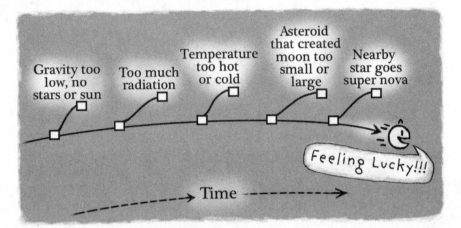

Fig. 10.2 Evolution of the universe as perceived by a conscious observer. The chain of "lucky coincidences" that leads to the observer experiencing the universe is one of the branches of the superposed wave function, collapsed relative to that observer.

In other words, the universe as experienced from the perspective of an observer is his consciousness. What an observer perceives as the external world is described in physics by wave function; wave function is a representation of an observer's awareness of the universe, not directly of the universe itself, which in fact does not exist without consciousness. You might stroll through a field, noticing the wildflowers—brilliant yellow, red, and iridescent purple. This colorful world constitutes your reality. Of course, to a mouse or a dog, that world of reds, greens, and blues doesn't exist any more than the ultraviolet and infrared world experienced by bees and snakes does for you. As we've seen throughout this book, reality isn't a hard, cold thing, but an active process that involves our consciousness. Space and time are simply the tools our mind uses to weave information together into a coherent experience—they are the language of consciousness. Regardless of differences in perception (many of which were discussed in the previous chapter), we genome-based creatures all share a common biological information-processing ability that allows us to arrange those perceptions into a spatiotemporal reality. "It will remain remarkable," said Nobel physicist Eugene Wigner, referring to a long list of scientific experiments, "that the very study of the external world led to the conclusion that the content of the consciousness is an ultimate reality."

The wave function, the multiverse, the realization that branching possibilities forever carry the living cosmos forward, and especially the final ingredient of the conscious observer all inescapably lead to a non-cessation of conscious experience. When we die, we do so within a matrix of inescapable life. Life transcends our ordinary linear way of thinking, even if we're handicapped by our ability to perceive only our current "world"—our single branch.

So what's it like when you die? In an article, Lanza offered a metaphor for the closing of one life chapter, and we'll use it here to close out this book chapter:

> During our lives, we all grow attached to the people we know and love, and cannot imagine a time without them. I subscribe to Netflix, and a few years ago worked my way through all nine seasons of the TV series Smallville. I watched two or three episodes every night, day after day, for months. I watched Clark Kent go through all the usual growing pains of adolescence, young love, and family dramas. He, his adoptive mother Martha Kent, and the show's other characters became part of my own life. Night after night I watched Clark use his emerging superpowers to fight crime as he matured, through high school and then college. I watched him fall in love with Lana Lang, and become enemies with his once friend Lex Luthor. When I finished the last episode, it was like these people had all died—the story of their world was over.
>
> Despite my sense of loss, I reluctantly tried a few other series, eventually landing on Grey's Anatomy. The cycle started over again, with completely different people. By the time I had finished all seven seasons, Meredith Grey and her fellow doctors at Seattle Grace Hospital had replaced Clark Kent, et al. as the center of my world. I became completely caught up in the swirl of their personal and professional passions.
>
> In a very real sense, death within the multiverse described by biocentrism is much like finishing a good TV series, whether it's Grey's Anatomy, Smallville, or Dallas, except the multiverse has a much bigger collection of shows than Netflix. At death, you change reference points.

It's still you, but you experience different lives, different friends, and even different worlds. You'll even get to watch some remakes—perhaps in one, you'll get that dream wedding dress you always wanted, or a doctor will have cured the disease that, in this life, shortened your loved one's time on Earth.

At death there's a break in our linear stream of consciousness, and thus a break in the linear connection of times and places, but biocentrism suggests that consciousness is manifold and encompasses many such branches of possibility. Death doesn't truly exist in any of these; all branches exist simultaneously, and continue existing regardless of what happens in any of them. The "me" feeling is energy operating in the brain. But energy never dies; it cannot be destroyed.

The story goes on even after JR gets shot. Our linear perception of time means nothing to nature.

As for me, as my own life's wave function collapses, I know that I'll still have season eight of Grey's Anatomy *to look forward to.*

THE ARROW OF TIME

Time is of your own making;
Its clock ticks in your head.
—Angelus Silesius

very story—including the epic narratives of our own lives—needs a framework, a skeleton. And every *exciting* story needs a villain. Time fulfills both requirements. For surely something must be blamed for the tragedy that transforms the beauty and vitality of our youth into the crepey skin and creaky joints of our aging selves.

This unspeakable crime's perpetrator was long considered an actual entity. Even great minds like those of Newton considered time a stolid feature of reality, an actual dimension through which all else passes. That notion of time as an absolute *thing*, ticking away outside of us, has never fully left the public mind. In 2014's sci-fi hit *Lucy*—in which the title character, played by Scarlett Johansson, is able to transcend many physical and mental constraints after being dosed with a drug—the brilliant scientist (played by Morgan Freeman) grandly informs us at the film's climax that time alone is real.

But the screenplay's writer could not have based that pronouncement on anything found in a modern physics text. Indeed, time's lack of reality is in some sense old news, going back at least to the head-spinning revelations of relativity.

According to Einstein's theory of relativity, there exists a four-dimensional continuum, with three spatial dimensions and one additional dimension, called "time." This connectedness between spatial dimensions and the temporal component threw most people for a loop. That's because in daily life, time seems utterly distinct from the three spatial realms. To review these three spatial realms via basic geometry, lines are one-dimensional; flat shapes like squares and triangles have two dimensions; and a solid form like a sphere or cube has three. However, an actual object—a sphere like an orange, say—requires an additional dimension because it persists and perhaps even changes. This means that something "else" besides its spatial coordinates is part of its existence, and we call this "time." This four-dimensional spacetime continuum is often referred to as "block universe," and it contains every possible point in space and in time, meaning that everything in it exists simultaneously—in the case of our four-dimensional orange, the various moments of its existence from ripe to rotten are all points in spacetime. There is nothing like the subjective experience of "becoming" or the sense of events unfolding in a temporal order.

Einstein, along with many scientists and philosophers, considered consciousness an extra ingredient, not belonging to traditional physics and the world it describes, and thus consciousness is not a part of spacetime, but moves through it. An observer's consciousness crawls along a line in the block universe. This line, called "world line," extends from the observer's birth to death.

So, little known to the general public, the word "time" has a double meaning. The time of Einstein's relativity, as explained above, is "coordinate time," one of the spacetime dimensions. If we talk about the year when Columbus discovered America, or an appointment with our boss one week ago, or any past or foreseeable future event, we have in mind the coordinate time of an event in spacetime. The event or point encompasses the time and place of the meeting with the boss, or the time and the street corner at which we caught a bus. Coordinate time does not move; each moment is a point existing within spacetime.

But in our everyday experience, "time" is anything but static—it is an unstoppable flow. When most people talk about time, this is the time they mean: a sequence of events that changes from moment to moment in our awareness. This is "evolution time," time as experienced by consciousness, the ever-changing "now."

For Einstein, such time was imaginary. When in 1955 he learned of his lifelong friend Michele Besso's death, he famously wrote to Besso's family, "Now he has departed from this strange world a little ahead of me. That means nothing. People like us, who believe in physics, know that the distinction between past, present, and future is only a stubbornly persistent illusion."

Einstein illustrated the relative nature of the perception of time through one of his famous thought experiments. Imagine you are sitting in the middle of a train while your friend is standing on the embankment outside, watching the train go roaring by. If lightning strikes on both ends of the train just as the train's midpoint is passing the embankment, your friend would see the two bolts of lightning strike at the same time, because both strikes are the same distance from him, the observer. If asked, he would say the strikes happened simultaneously—an accurate statement of his perception of time. However, from your perspective, sitting in the middle of the train as it moves forward, you will see the lightning that strikes the front of the train first because the light of the lightning in the rear has a slightly greater distance to travel to you. As a result, if asked, you would say the lightning strikes were not simultaneous, and that the one in front actually happened first—an accurate statement of *your* perception of time. In this and other thought experiments, Einstein showed that time actually moves differently for someone in motion than for someone at rest, and it only exists relative to each observer. In the case of the train and the lightning, neither your observation nor your friend's is "more right"—there is no objective viewpoint, just two different perceptions.

Biocentrism takes this a step further by suggesting that the observer doesn't just perceive time, but literally creates it. Most people take for granted the reality of what our minds put together. We understand dreams as a mental construct, but when it comes to the life we live, we accept our perception of time and space as absolutely real. But in fact, as we've seen throughout this book, space and time are not objects. Time is simply the ordered construction of what we observe in space—much like the frames of a film—occurring inside the mind.

According to biocentrism, these mental constructs are based on algorithms, or complex mathematical relationships, whose physical logic is contained in the neurocircuitry of the brain. The particular algorithms your brain uses to translate the welter of perceptions flooding your senses into a coherent, lived experience are the key to consciousness, and they also explain why time and space—indeed, the properties of matter itself—are relative to the observer.

In the end, life is motion and change, and both are only possible through the representation of time. At each moment we are at the edge of a paradox known as "the arrow," first described 2,500 years ago by Zeno of Elea. Since nothing can be in two places at once, he reasoned that, during any given instant of its flight, an arrow is in only one location. Thus, at every moment of its trajectory, the arrow must be present at some specific place. But if the arrow is in one specific place, it follows that it must momentarily be at rest. Logically, then, as the arrow flies from bow to target, what is occurring is not motion per se, but a series of separate static events. The forward motion of time, embodied by the movement of an arrow, is not a feature of the external world but a projection of something within us that ties together things we are observing.

In 2016, one of the authors (Lanza) published a scientific paper with Dmitriy Podolskiy, a theoretical physicist then working at Harvard. The paper appeared as the cover story in *Annalen der Physik*—coincidentally, the same journal that published Einstein's theories of special and general relativity. It explains how the arrow of time, and time itself, emerges directly from the observer: that is, from us. Time, it argues, does not exist "out there," ticking away from past to future, but rather is an emergent property that depends on an observer's ability to preserve information about experienced events.

Time is inarguably a relational concept—one event relative to another. Time as we experience it has no meaning without association to another point. Thus, it requires an observer with memory; without such an observer, one cannot have the relational concept that lies at the core of any "arrow of time."

Fig. 11.1 Podolskiy and Lanza's paper on the arrow of time was the cover story of Annalen der Physik, *which had published Einstein's theories of special and general relativity. In his papers on relativity, Einstein showed that time was relative to the observer. This new paper (reproduced in its entirety at the end of this book) takes this one step further, arguing that the observer creates it. The arrow of time depends on the properties of the observer, and in particular, the way we process and remember information.**

* D. Podolskiy and R. Lanza: *Annalen der Physik* 528 (9–10), 663–676, 2016.
Copyright Wiley-VCH Verlag GmbH & Co. KGaA. Reproduced with permission.

Time as an arrow is a metaphor that dates back millennia. The expression "arrow of time" arises because time as we experience it displays a directionality, an ability to change in one way but not the reverse. A car's metal may crinkle and bend in a fender bender, but a vehicle thus ruined cannot move backward while its material irons itself out to become unscathed again. To early iconographers, no item so perfectly represented this feature and limitation of time—its status as a strict one-way street—as a flying arrow. They might have chosen the image of a horse or fish facing in some particular direction since it's rare to observe either of them propelling themselves backward, tail first. But rare is not impossible. Other natural objects were even more problematic; lightning can go from cloud to ground or vice versa, and anyway, how would you visually distinguish its front end from its caboose? Happily, arrows are infallibly one-way devices, their point always leading the way; even if one is shot straight up and then starts backsliding, it quickly flips over.

But is time—evolution time, the arrow-like time we experience—an idea or an actuality?

The reality of time appears indispensable to anything involving change, such as the growth of a stalactite in a cave, even if the process of change in that case is so slow as to take five hundred years for the buildup of a single inch. But where generations of physicists have been most likely to point as evidence of the reality of time is the second law of thermodynamics, which describes *entropy*—the process of going from greater to lesser structure and order, like what happens to your underwear drawer over the course of a week.

Consider a glass of club soda over ice. At first, everyone can see that there is definite structure. The ice cubes float at the top and keep themselves a bit separate from the liquid. Bubbles come and go. The ice and seltzer have different temperatures. But return later and you'll find the soda has gone flat, the ice has all melted, and the contents of the glass have become a structureless sameness. With no temperature variations, energy transfers have more or less stopped at the atomic level. The party seems like it's over, because it *is* over. Barring evaporation, nothing further will happen.

This evolution away from structure, order, and activity toward uniformity, randomness, and inertness is known as the increase of entropy. One of the most basic and important concepts in physics, it is a process that pervades the universe and may even call the shots cosmologically in the long run. Today we see individual hot spots like our sun spewing heat and ions into their cold environs, but this organization is slowly dissolving.

This overall loss of structure is a one-way process. As described by the second law of thermodynamics, the increase of entropy—like the crumpling of a car in a fender bender—is a nonreversible mechanism. Thus, it doesn't make sense without a directionality of time. In fact, it *defines* the arrow of time. Without entropy, it need not exist at all.

A fascinating point is that, despite years of physicists pointing to entropy as proving the existence of time, Ludwig Boltzmann himself—the scientist who actually discovered and developed the three laws of thermodynamics—didn't share this view.

Using the scrupulous logic characterizing his field of statistical mechanics, Boltzmann insisted that increasing (and never decreasing) entropy is simply what we must observe by virtue of living in a world where disordered states are the most probable. A dynamically ordered state, one with molecules moving "at the same speed and in the same direction," Boltzmann concluded, is "the most improbable case conceivable . . . an infinitely improbable configuration of energy." Imagine being handed a deck of cards, with all numbers strictly sequential and each suit separated from the others, as if fresh out of its box. Could you ever be convinced that this was a shuffled deck, one that just happened to have arranged itself in that particular order rather than any of the other possible configurations? Because there are so many more possible disordered conditions than ordered ones, a state of maximum disorder is simply the most likely to appear. The fact is, order anywhere in the cosmos is so singular, it always requires an explanatory mechanism or process, while randomization requires no elucidation—it is simply the way of the world. When particles are allowed to act randomly, whether as the contents of a glass of club soda on the rocks or the countless atoms making up the air in a room, they slam into each other, exchanging energy until complete randomization of their positions and velocities is observed.

So no "arrow" need exist. Entropy is merely an outgrowth of ordinary random behavior. The second law of thermodynamics, stating that entropy never decreases, is an automatic consequence of statistical probability. It requires no external entity dictating its direction.

Similarly, the brilliant scientists who developed the most fundamental explanations of the functioning of our natural world—Newton's laws of motion, special and general relativity, and quantum theory—found that their equations all function independently of the notion of the passage of time! They are "time symmetrical," meaning they operate backward as easily as forward. The arrow of time has no place in them.[†]

Metaphysicians, taking entirely different routes, have also questioned time's reality. The past, they say, is just an idea in a person's mind; it is no more than a collection of thoughts, each of which is only occurring in the present moment. The future is similarly nothing more than a mental construct, an anticipation. Thinking itself occurs strictly in the "now"— so where is any flow of time?

Statistics, equations, and metaphysics aside, it shouldn't come as too much of a surprise that time has no existence as an independent entity, external to us observers. For who except observers experiences any change? As we've seen, without observers, reality doesn't exist to begin with—but it's doubly sure that it couldn't exist as a series of events threaded together as a linear unfolding.

Nevertheless, we creatures of consciousness experience time as a seemingly relentless forward march. Humans have long been fascinated by this march, and our imagination captured by imagining its reversal. Any one-way process must naturally conjure bizarre consequences if it were somehow made to run in the "wrong" direction. Like time, gravity is a one-way phenomenon in that it pulls but never pushes. This strict directionality of its force is so baked in to human experience that

---

† The "time" in those equations is not evolution time, it is just coordinate time. In general relativity, even time as a coordinate loses its role, which leads to the famous "problem of time" in quantum gravity, which we'll discuss in Chapter 14.

sci-fi easily denotes weirdness with images such as water spiraling up out of a drain, and in the late fifties, NASA was convinced that violating or nullifying gravity's pull might have grave or even lethal human consequences—which is why they launched chimpanzees in test flights before they dared send astronauts into space. The potential effects of reversing time's arrow have been the source of much scientific debate, for instance about whether effect could ever precede cause, and what that might mean.

As we noted at the beginning of this chapter, the arrow of time is often cast as life's villain, guilty of the crime of robbing us of youth; the theme of defeating it was explored in the 2008 movie *The Curious Case of Benjamin Button*, which was based on a 1922 short story of the same name. In the movie, Brad Pitt springs into being as an elderly man and ages in reverse. The movie was massively popular, and it got many people thinking about time's arrow and its implications.

We've seen that, to the bafflement of scientists, the fundamental laws of physics have no preference for a direction in time and work just as well for events going backward as they do going forward. Yet, in the real world, coffee cools and cars break down. No matter how many times you look in the mirror, you'll never see yourself grow younger. We're left with a major contradiction between what we ourselves routinely experience and what science says must be true. If time is an illusion, you might be wondering, why do we age? If the laws of physics should work just as well in either direction, why do we only experience growing older?

The answer, once again, lies in us observers—specifically, in our function of memory. If time is truly symmetrical in equations stretching from Newton to modern quantum mechanics, then science would seem to say that we should be able to "remember" the future just as we experience the past. But quantum mechanical trajectories "future to past" would be associated with the erasure of memories, a decrease in entropy leading to a decrease of entanglement between our memory and observed events. Hence you can't go back in time without information being erased from your brain; if we do experience the future, we are not able to store memories about such experiences to "remember" back in our present. By contrast, if you experience the future by traveling the usual one-way path, "past > present > future," the random process of entropy continues, and you continue to only accumulate memories.

So aging, too, offers no proof of time's arrow as an external force. It seems that time truly does not exist outside of awareness; it is consciousness itself, by being accompanied by mechanisms like memory that allow for comparisons, that ushers in the arising of time as surely as the sunrise dispels the night.

In the world of biocentrism, a "brainless" observer—that is, one without the ability to store memory of observed events—does not experience a world in which we age. But it goes further and deeper than this: a brainless observer does not merely fail to *experience* time—time does not exist for such an observer in any sense. Without a conscious observer, the arrow of time—indeed, time itself—simply doesn't come into existence in the first place.

In other words, aging truly is all in your head.

TRAVELING IN A TIMELESS UNIVERSE 12

Time and space are but physiological colors which the eye makes.
—**Ralph Waldo Emerson**

We are all time travelers. From waking up in the morning to going to sleep in the evening. From arriving on the job at 9 AM to finishing at 5 PM. From leaving for vacation in late August to returning home two weeks later to find the first hint of fall in the air. We travel in time from birth to death.

In television's joyful cult classic *Doctor Who*, a two-thousand-year-old "Time Lord" from the planet Gallifrey traverses both time and space in a craft called the TARDIS. While the TARDIS (an acronym for Time and Relative Dimension in Space) looks on the outside like an ordinary-if-old-fashioned British police telephone box, its vast interior is a technological marvel that has seemingly marshaled the laws of physics to enable visits to such temporally remote locales as 1814 London, the Jurassic period, and even future cities on distant planets.

But will *we* ever be able to travel back and forth in time like the Doctor? Could we build a chariot capable of transporting us around the universe not just in three dimensions, but in four? When we speak of "traveling in time" in this way, of course, we mean traveling along *coordinate* time, as distinct from the usual everyday traveling of consciousness from the morning to the evening and so on, proceeding ever forward along the path of our life. Such traveling of consciousness is often called the "passage of time," in spite of the fact that time does not pass—it is our awareness that passes along the time coordinate.

In classical science, humans have placed all things in time on a linear continuum. The universe is nearly fourteen billion years old, the Earth around four to five billion years old, and we ourselves twenty or forty-five or ninety years of age. In the common conception of an external mechanistic universe, time is a clock that ticks independently of us. Not so, says biocentrism. As physicist Stephen Hawking pointed out, "There's no way to remove the observer—us—from our perceptions of the world." The world we perceive is created by us. And Hawking believed that it wasn't just our present reality that we create, but rather that the universe similarly has many possible histories and possible futures. Remember: "In classical physics," he said, "the past is assumed to exist as a definite series of events, but according to quantum physics, the past, like the future, is indefinite and exists only as a spectrum of possibilities."

We've been taught that our consciousness—and everything else in the world—flows like an arrow in one direction from the cradle to the grave. But in the previous chapter, we saw that this arrow is not something external to our consciousness; it is only created by it. And an amazing set of experiments suggests the past, present, and future are entangled—and that decisions you make now may influence events in the past.

We are referring, of course, to "delayed-choice" experiments of the sort discussed in Chapter 7. As you'll recall, they were first envisioned by our friend Wheeler, and in 2007 such an experiment was at last carried out and published in the journal *Science*. You should feel free to flip back to Chapter 7 for a second look at the illustration and details of the setup, but in short, the scientists shot photons into an apparatus and showed that they could retroactively alter whether these photons behaved as particles or as waves. When they passed a fork in the apparatus, the photons had to "decide" what to do—but later in the photons' journey, after they'd traveled nearly fifty meters past the fork, a switch flipped by the experimenter could determine how the photon had behaved at the fork in the past.

The results of this experiment and others like it were no small revelation. And for most of us, it may take a while to fully grasp that the past is not inviolable; that, like the future, it is determined by current events.

What's more, following this logic leads to a further conclusion, namely that what happened in the past may depend not only upon decisions you make now—it may even depend on actions you haven't taken yet.

According to Wheeler, "The quantum principle shows that there is a sense in which what an observer will do in the future defines what happens in the past." Quantum physics tells us that objects exist in a suspended state until observed, at which point they collapse into a definite reality. Wheeler insisted that, when observing light bent around a galaxy from a distant quasar, we have in fact set up a quantum observation on an enormously large scale. In other words, he said, the measurements made on incoming light now determine the path that light took billions of years ago. This mirrors the results of the experiment described in Chapter 7, where a present observation determines what a particle's twin did in the past.

In 2002, *Discover* magazine sent a reporter to Maine to speak to Wheeler firsthand. Wheeler said he was sure the universe was filled with "huge clouds of uncertainty" that haven't yet interacted with anything. In all these places, he said, the cosmos is "a vast arena containing realms where the past is not yet the past."

There remains a fluidity—a certain degree of uncertainty—to anything that is not actually observed. Part of the past is locked in when you observe the world around you in the present and probability waves collapse. But there's still some uncertainty—for instance, as to what's underneath your feet. Before you observe what's there, the particles that make up what's underneath you have a range of possible states, and it's not until observed that they take on real properties. So until the present is determined, how can there be a past? If you dig a hole, there's some probability that you'll find a boulder. If you do, the glacial movements of the past that account for the rock being in exactly that spot will then coalesce into certainty. The past is simply the spatiotemporal logic of the present, including the geological history that corresponds to the branch of reality your consciousness collapses.

Bottom line: reality begins and ends with the observer, whether you're speaking of the reality of now or that of an eon ago. "We are participators," Wheeler said, "in bringing about something of the universe in the distant past."

Like the boulder in your backyard and the light from Wheeler's quasar, historical events such as who killed JFK might also depend on events that haven't yet occurred. You only possess fragments of information about the event; there's enough uncertainty that it could be one person in one set of circumstances or another person in another. History is a *bio*logical phenomenon. It's the logic of what you, the animal observer, experience. You have multiple possible futures, each with a different history. As you live, observing and acquiring information, you collapse more and more reality. Perhaps choices you make today will influence events far before your birth, rearranging the reality of occurrences from the time Christ was born, or when the Great Pyramids were being built.

Aristotle obviously failed to anticipate quantum mechanics when he said, "This only is denied to God: the power to undo the past."

But altering the past is one thing—can we ever hope to travel to it?

We live and die in the world of here and now. But this could change once science has a full understanding of the algorithms we employ to construct the reality of time and space. Although time does not exist per se, travel into past and future universes is likely possible if we are able to generate consciousness-based reality. If we then changed the algorithms—so that instead of time being linear, it was three-dimensional, like space—consciousness would be able to move through the multiverse. (A vivid illustration of what such traveling through the multiverse might look like is given in the fascinating science fiction novel *The Other Side of Time*, by Keith Laumer.)

Various theories involving additional dimensions of time have been considered in the scientific literature. However, the prevailing consensus has been that multidimensional time is impossible, because by implying the possibility of motion into the past, it gives rise to causal paradoxes—and it is taken for granted that any theory leading to causal paradoxes is not physically viable and must therefore be rejected. Such was the fate of the theory of tachyons (particles that move faster than light). Special relativity can be extended to incorporate superluminal velocities, but curiously, the formulation of the resulting extended relativity only works if one postulates that not only space, but also time, is three-dimensional.

So while three-dimensional time would allow for time travel, it is widely considered untenable for just this reason, as time travel would itself allow for contradictions such as the classic "grandfather paradox." In this well-known example, a person travels back in time and kills their own grandfather before their mother or father's conception. Thus the traveler would never have been born, and would not have been able to travel to the past to kill their grandfather in the first place.

There are many other equivalent paradoxes and inconsistencies that might emerge through changing the past, such as the (im)possibility of going back in time and killing oneself as a baby, or the famous "Hitler paradox," in which killing Adolf Hitler erases your very reason for going back in time to kill him. Paradox aside, killing Hitler in the past would have monumental consequences for everyone in the world today, especially for those who were born after World War II and the Holocaust. If you'd killed Hitler, none of his actions would have trickled down through the subsequent years, both for better and for worse. Millions of people who would have died would now have lived, but there would be countless other changes difficult to predict: people who met and had children might never have known each other, whole nations might exist in different forms or not at all, not only the atomic bomb but all sorts of other technology might never have been invented. The entire course of history would have been unrecognizably different.

This problem is explored in an episode of *Doctor Who* aptly called "Let's Kill Hitler." The TARDIS crash-lands in Nazi Germany just as a humanoid robot is about to kill Hitler, and the Doctor and his companion proceed to save Hitler in the past in order to likewise save their future. Similar dilemmas inform the plotlines of *The Terminator* and

Back to the Future movies, where visiting the past constantly threatens to rearrange the future from which the travelers came.

Despite ingenious attempts to get around these blockades, timeline inconsistencies are indeed problematic for time travel as it is classically conceived. But all of these paradoxes disappear if the rules of quantum mechanics apply to the macroworld, with no single past and multiple possible futures. According to the many-worlds interpretation, if you traveled back in time, you would simply create an alternative timeline or a parallel universe. Whether flipping a switch (like the scientists conducting the delayed-choice experiment described in Chapter 7), or turning the dial of a time machine, it is always still you inside the experience. There can be no paradoxes because any event you alter in the past will generate an alternate universe in keeping with the known laws of quantum mechanics. No matter which universe you inhabit, you inhabit it as yourself.

Of course, time travel going toward the *future* is another matter entirely, and—as it avoids pesky paradoxes like those described above—its theoretical mechanisms are relatively straightforward even in classical physics. We know from Einstein's theory of special relativity that time passes at different rates depending upon how fast objects are moving. This "time dilation" becomes huge as you near the speed of light. For instance, for someone traveling at about 580 million miles per hour, a clock would run half as fast as it does for someone at rest. Thus, to travel quickly ahead in time—that is, without aging excessively on the way there—you'd just need to travel near light speed for a bit, and then turn around and come "back to the future" you were aiming for.

However, while (with the right equipment) traveling to the future in this way is theoretically possible, there are a few, er, "minor" snags. For instance, Einstein showed that nothing that weighs anything can quite attain the speed of light, because its mass would grow until, at just below light speed, even a feather would outweigh a galaxy. The amount of force needed to accelerate such a now-huge mass further, to light speed, would be impossible to obtain—it would exceed all the energy in the universe. Indeed, at just below the speed of light, a zooming mustard seed would outweigh the entire cosmos.

Forward time travel might also theoretically be achieved by exploiting the properties of gravity. Einstein's theory of general relativity tells us that it isn't only motion that affects the speed of time; clocks also tick more slowly in stronger gravitational fields. A clock on Earth, like that at Mission Control in Houston, ticks a tiny bit slower than a clock on the moon. In fact, there are places in the universe where only a single second of time passes while a million years' worth of events simultaneously unfolds here on Earth. Unfortunately, traveling any significant distance in time by way of gravitational time dilation would require extreme (and, alas, likely deadly) measures, such as orbiting close to a black hole at tremendous velocities or traveling to a neutron star. Of course, to do the latter, you'd need a machine with a spherical shell weighing a million times more than the earth. Standing on a neutron star—even if you could build a starship to get there—would flatten you as effectively as one of the huge boulders that falls on Wile E. Coyote in his pursuit of the Road Runner.

Some of the more well-known theoretical time travel possibilities proposed by scientists involve strange configurations of spacetime such as "wormholes," oddities that contain so-called "closed time-like loops" that would allow a particle to travel back in time and meet itself. However, while the equations of general relativity allow for such things, construction of wormholes is not possible without exotic theoretical materials that have not been found in nature, at least not yet. And of course, in most of these theories, there is no way a traveler can go back to a time before the "time machine" itself was built.

To sum up, building a machine to travel through time like the Doctor in *Doctor Who* is impossible via classical physics, whether because of causal paradoxes or practical difficulties. The findings of quantum theory suggest both intriguing solutions to some of these problems, and that the past and future themselves are not the definite and separate realities they seem. But it is when we fully incorporate the principles of biocentrism, shifting our worldview by adding life to the equation, that things really get interesting. By accepting space and time as forms of animal understanding (that is, as *bio*logical) rather than as external physical objects, we may open a completely new vista for time travel.

We have seen that in the multiverse of many possible histories and parallel universes, a setup in which causal paradoxes simply do not exist, time travel may well be possible. But the very word "travel" implies motion to a place that is distinct and separated from ourselves, with the challenge of physically shifting mass (our bodies) and minds (our consciousness) to a new position in space and time. What if time travel is found not to require a displacement to somewhere "over there," but rather the mere experience of another aspect of "right here"?

According to biocentrism, space and time are relative to the individual observer—we carry them around as turtles do their shells. If you accept that neither space nor time has any stand-alone existence, that both are instead inseparable functions of the algorithms comprising our consciousness, it should be obvious that "travel" through either dimension may not ultimately entail any kind of physical journey at all.

As technology exploits the new biocentric paradigm, then, time travel may well prove to be far easier than it sounds.

THE FORCES
OF NATURE

13

*The universe is the externalization of the soul. Wherever
the life is, that bursts into appearance around it.*
—Ralph Waldo Emerson

Amazing "coincidences" confront us the moment we ponder the universe. But they morph from inexplicable oddities to profound revelations once we fully grasp the intimate connection between the seemingly vast and distant cosmos and our own minds.

We have said that the universe is an information system that is in fact nothing more or less than the spatiotemporal *logic* of the observer, meaning the self. This alone explains why the laws and forces of nature—which could have almost any value—are all exquisitely balanced in favor of our existence. It is why, for instance, the value of the strong nuclear force is within the narrow range that allows the atomic nuclei in our bodies to hold together without disastrously binding protons together as well. It explains why the gravitational force is exactly as it must be for the sun to ignite, and for fusion to proceed, generating the forces needed to make the carbon atoms that are the very backbone of life itself.

When Emerson asked, "Does not the eye of the human embryo predict the light?" he was perceiving this intimate tie-in. This is why trying to understand how the universe all goes together is, in a way, like trying to understand the algorithms in a calculator, except in this case, we want to understand the internal logic of our own mind, to grasp how its effortless unseen mechanisms construct the various building blocks of spatiotemporal reality.

Earlier in the book, we explored how consciousness works, starting with ion dynamics at the quantum level within the neurocircuitry of the brain, and how that process of consciousness collapses the physical world we observe. Since reality is observer dependent, the spatiotemporal machinery by which consciousness manifests as real-life, 3-D objects and events can in fact be extrapolated in space from the quantum realm to the edges of the universe, and in time until the footsteps of our ancestors disappear into the sea.

Of course, cosmologists have picked up the story of the molten Earth and carried its evolution backward in time to the insensate past: from minerals by degrees back through the lower forms of matter—those of plasma and nuclei, of quarks—and beyond this to the big bang. Indeed, if we could travel back in time, we would probably observe most if not all of those events predicted by cosmologists. But, as we have seen, physical reality begins and ends with the observer. What is observed is real; all other times and places, all other objects and events are products of the imagination and serve only to unite knowledge into a logical whole. Think of the universe as a globe you might see in a classroom—it's merely a representation of everything that's theoretically possible to experience (assuming, of course, we were able to get there and survive long enough to observe it).

One of the goals of this chapter is to untangle the logic that the mind uses to generate such a spatiotemporal experience. While experiencing consciousness, the mind employs an algorithm, a mathematical rule that provides the precise relational logic to define and animate the construction. We might start with the logic of the electromagnetic wave, which defines the interrelationship of space and time in a precise mathematical way. Let's jump right in.

In his seminal paper "On the Electrodynamics of Moving Bodies," Einstein discovered how to make sense of the discrepancy between how material bodies and electromagnetic waves move. He constructed his special theory of relativity, which unified on the one hand space and time, and on the other hand matter and energy. Einstein's findings condense into the famous formula $E = mc^2$, which says that whatever units are inserted (for instance, ergs for energy, grams for mass, and c expressing light speed in centimeters per second), the energy of an object will be found to be exactly equal to its mass multiplied by the square of light's velocity. That this was not merely mathematically elegant but also perfectly correct was vividly illuminated during World War II—by the fireball of the first atomic bombs. By sheer coincidence, in both the Trinity and the Nagasaki devices, only a single gram of each bomb's fourteen pounds of plutonium was converted to energy and vanished. Yet this single gram was enough to create a titanic explosion dwarfing anything the world had ever seen. It was also a powerful physics demonstration: substitute the number "1" for the m in the $E = mc^2$ equation, and it will tell you that a single gram should convert to the energy equivalent of 21,000 tons of TNT—exactly the yield of the Nagasaki bomb.

Einstein's unification of the equations that describe electromagnetism (the so-called Maxwell equations) and the motion of matter required introduction of a four-dimensional continuum, combining space and time.

Taking Einstein's postulates to their rigorous mathematical conclusion, Hermann Minkowski introduced the concept of *spacetime*, a four-dimensional space whose points need four numbers to be fully specified—three spatial coordinates, and one for time as well. When formulated within such a four-dimensional framework, the joint theory of light and matter became consistent.

So, each event in spacetime is described by three coordinates denoting its spatial position plus the additional coordinate called "time." Thus, when you arrange an appointment with somebody, you specify not only a place, but also a time of the event.

But alas, in naming the fourth coordinate "time," Einstein and Minkowski used a concept that is traditionally linked to our subjective feeling of "becoming," namely, our experience of events as occurring one after the other, sequentially. But, as explained in Chapter 11, our subjective "time" is not the same time as the fourth coordinate of the spacetime continuum—a source of eternal confusion among laypeople trying to understand Einstein.

You'll remember that in the "block universe" of the spacetime continuum, everything exists simultaneously; there are no dynamics, no subjective experience of "becoming" or events unfolding in a temporal order. Thus, the block universe of special relativity is of course not consistent with what humans actually observe. We do not observe past, present, and future all at once. We observe time as unfolding in our consciousness piece by piece, event by event. Physicists consider such unfolding of time an "illusion," something that happens only in consciousness, which for them is not a part of physics. And yet, this formulation, without their being aware of it, hits at the most basic fact of existence.

To elaborate a bit, the word "illusion"—which arises repeatedly in this chapter—is actually denoting the fact that the involvement of consciousness is deeply intertwined with the workings of the universe, and that the block universe is not enough. For the science to work, it turns out one needs an extra ingredient, and this seemingly spooky element that was not an established part of physics was consciousness.

Early in the twentieth century, physicists were dimly aware of this, though it became increasingly clear with the advent of quantum mechanics, which doesn't make much sense unless one brings consciousness into the game. Acceptance of this fact still sees reluctance among scientists, even now, a century later. Apparently, our scientific culture is as resistant to radical changes of its paradigms as were the contemporaries of Copernicus and Galileo.

According to quantum mechanics, not only the unfolding of time but the existence of external events, and thus the universe as a whole, is in a sense merely an illusion of the observer. As biocentrism emphasizes, the outcome of an experiment or an observation is awareness in the mind of the observer. In a real sense, the word "external" is a hollow term, since nothing is external to consciousness—or the observer's mind.

So "illusion" arises in several ways. The "time" utilized by special relativity as the fourth coordinate of the spacetime continuum is in some sense illusory, too, since as we've seen it is not true time at all. It is "time" only in the way the position of a clock's hands depict time. A moving indicator on a clock face can have any position, but all such configurations exist simultaneously in spacetime. It is consciousness that determines which position the clock's hands are assuming right now. This shortcoming—special relativity needing some sort of modification to introduce an extra parameter that would account for our subjective experience of "becoming"—has been realized and thoroughly investigated by many physicists, starting with an initial proposal by Ernst Stueckelberg. Lawrence Horwitz introduced the distinction we discussed in Chapter 11, between the *coordinate time* that constitutes the fourth dimension of spacetime, and the *evolution time*, associated with an extra parameter.

As we said above, the points of spacetime are associated with events. But what are these, exactly? An event can be the place and the time where and when a particle hits another particle. For instance, a photon scatters from an atom and arrives into the observer's eye, thus bringing the observer information about the atom's position. But according to quantum theory, the atom's position is "blurred" because it is a superposition of many possible positions. It is the job of consciousness to determine which of those possible positions becomes the actual position in the observer's awareness. By the act of observation, the observer's mind creates the awareness that, at a certain position and at a certain *coordinate time*, the photon scattered from the atom. Another photon may arrive, followed by yet another. The cascade of events in spacetime corresponds to a cascade of experiences in the observer's consciousness. But without a mechanism in the mind to sort these events successively in order to experience them one by one, all those experiences would exist simultaneously: past, present, and future.

A coherent state of many photons manifests itself at the macroscopic scale as an electromagnetic field. This electromagnetic field may take the form of electromagnetic waves such as radio waves, infrared light, and so on. Electromagnetic waves whose crests are anywhere between 400 and 700 nanometers apart constitute visible light, and are our primary tool of investigating our surroundings. And by surveying the positions of objects around us, we are able to envisage, piece by piece, the entire universe as existing in spacetime.

In a sense, waves of electromagnetic energy are the fibers of logic the mind uses to weave a four-dimensional tapestry. It is a mathematical relationship that not only defines the four dimensions of spacetime, but that also defines how *evolution time* is infused into that spatial construct—it is the logic that generates the experience we call "motion." Through the incorporation of memories, the mind uses this logic to generate the complex information system that we experience as consciousness or reality. The fact of simple everyday awareness involves underlying mechanisms of astonishing intricacy. It is an act that borders on magic, considering the countless "alterations" that might theoretically occur at each point in three-dimensional space.

And just as we can magnify our senses with radio telescopes to see through the opaque dust clouds of the Milky Way galaxy, we have scientific tools that enable us to analyze what's going on at the unseen heart of all physical events. In conformity with Maxwell's equations, it turns out that the electric and magnetic components of an electromagnetic wave stand together in time relationships, each depending on the other's rate of change. The mutating electric field at one point in space generates a magnetic field at right angles to it, which generates a follow-up electric field, and so forth, with the process propagating to limitless distances at the speed of light—an unvarying 186,282 miles per second (see figure 13.1).

When we stand face-to-face with an object, we see light glimmer on all its surfaces as if it were an independent object, standing outside and apart from us. No microscope could find the umbilical cord connecting the object to the mind of its observer, yet shape, sound, motion, resistance—all these things are nothing but energy impressed upon our sensory organs. But despite attempts to define or explain this energy, there has always remained in the last analysis of our experiments this residuum we could not resolve. It is an enigma whose source is hidden. Indeed, when Einstein

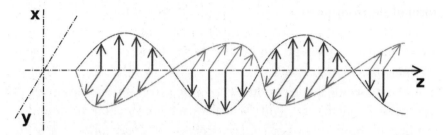

Fig. 13.1 An electromagnetic wave, propagating in the direction +z through a vacuum. The electric field (bold arrows) oscillates in the ±x-direction, and the magnetic field (gray arrows) oscillates in phase with the electric field, but in the ±y-direction.

was asked by David Ben-Gurion if he believed in God, he responded to the effect that "there must be something behind the energy."

This something need not, however, be sought in the material world. Energy is but a representation of the mind, a rule of its understanding. In the mind, were we able to lay it open, we would see the internal logic of the universe. It is here that sensations pass into appearance, and the components of electromagnetic waves generate the spatiotemporal relations necessary for us to understand and experience the empirical content of the physical world. Only by such a form of understanding can we apprehend the continuity in the connection of times and spaces.

The answer lies not in any isolated, external definition of "nature," but in ourselves; the mind makes the body, as the poet Spenser said:

> *So every spirit, as it is most pure,*
> *And hath in it the more of heavenly light,*
> *So it the fairer body doth procure*
> *To habit in, and it more fairly dight,*
> *With cheerful grace and amiable sight,*
> *For, of the soul, the body form doth take,*
> *For soul is form, and doth the body make.*

In other words, this time, those of Emerson: "The universe is the externalization of the soul. Wherever the life is, that bursts into appearance around it." We must not seek the original laws and forces of physics in nature, but rather we must seek them in our own mind; in the way the brain, operating through the systems of the body, generates knowledge of the sense environment.

The material, spatial world has no less a root in the mind than the poems of Aeschylus or Ovid. When we analyze the objects around us, we find nothing at last but energy—energy impressed upon our organs of sense, or energy resisting our organs of action. There is no object that cannot be resolved to this residuum. We are thus more than mere spectators of events. As is shown unambiguously in quantum physics experiments, the observer interacts with the system to such an extent that the system cannot be thought of as having an independent existence.

Our difficulty in apprehending this results from the impression that our awareness of our own existence is bound up with the objects around us. You go to the street corner, and with one glance at the morning paper, you can determine your location in time. Your eyes are bathed in lights and forms, your ears in the roar of cars and the chatter of pedestrians. You can establish your place in an instant. Yet this does not actually require anything self-subsistent or permanent. There must be, however, in the mind, a rule through which one state determines another, and also reverse-wise, an event's position in time and space. Read about the transformations of energy into matter. Find yourself in the laboratory, watching scientists create particle-antiparticle pairs from electromagnetic energy. Pass by the cloud chambers and watch the newly created matter leave in its wake thin, transient lines of white vapor. In the end, the unseen umbilical cord between mind and matter remains.

Emerson was right: "A man is a bundle of relations, a knot of roots, whose flower and fruitage is the world." It is startling to realize that not only are objects mere appearance, but that even their shapes are nothing but a form of the mind. Still, those objects we perceive around us are very different from our thoughts and feelings, from love and anxiety, joy and sorrow. Our thoughts and desires, the textures of our experience, can never be found among the atoms and objects of the outer world. And yet all these are also connected to each other through the time relations of electromagnetic energy, making the latter an entity that indeed unifies the mind with matter and the world.

Mind, matter—reality is a curious process. It is being continuously coordinated in your head. A moment can never pass without the mind gluing the past and the present together. You hear the telephone or the doorbell ring, but it cannot happen until the sound is actually in the past, until the mind compares it to the silence of a moment or two before. Even now, you cannot read this sentence until your mind compares the white here to the black there, now a letter, now a word, putting it all into some sort of contrasting order.

The fact is that both the temporal reality of unfolding events (in the sense of the evolution time mentioned above) and the spatial reality of the outer world exist only through an active exercise of the mind. They operate in perfect unison like a single timepiece.

What skill the mind shows in the fabrication of its own web! Think of the mind as attached to energy as effortlessly as barely-there threads of gossamer floating through calm autumn air, utilizing the electric and magnetic components that interact at intervals and define the space through which they are passing. And then marvel at this scaffolding, at how there is no known supporting structure below it; it is only a web of information floating above the void of nonbeing.

But electromagnetism is just one of several basic relationships—commonly called "forces" or "interactions"—that the mind uses to construct reality from all the possibilities implied by quantum mechanics. The other three fundamental interactions are the strong interaction, the weak interaction, and gravitation. We will not go into detail for each of them; suffice it to say that they, too, have their roots in the logic of how the various components of the information system interact with each other to create the 3-D experience we call consciousness or reality. Each force describes how bits of energy interact at different levels, starting from the foundation up—the strong and weak forces govern how particles hold together or fall apart within the nuclei of atoms, whereas electromagnetism and gravity enjoy infinite range, although the latter dominates the interactions on astronomical scales such as the behavior of solar systems and galaxies.

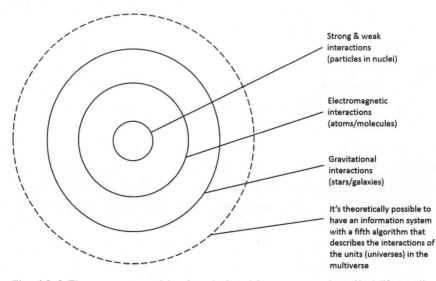

Strong & weak
interactions
(particles in nuclei)

Electromagnetic
interactions
(atoms/molecules)

Gravitational
interactions
(stars/galaxies)

It's theoretically possible to
have an information system
with a fifth algorithm that
describes the interactions of
the units (universes) in the
multiverse

Fig. 13.2 There are several basic relationships, commonly called "forces" or "interactions," that the mind uses to construct reality. Each force describes how bits of energy interact at different levels, starting with the strong and weak forces and moving up to electromagnetism and gravity. Theoretically it might be possible to also add another building block to the algorithms, one that governs the interactions of the units (universes) in the multiverse.

These are the algorithms that define our universe. Theoretically it might be possible to add another algorithm, one that governs the interactions of the units (universes) in the multiverse (see figure 13.2), this time in the sense of the inflationary scenario,* where our universe is just one of the bubble universes, each of the others containing a slightly different history from ours. For instance, you might be able to step into a room where your dead cat is still alive, or where 9/11 never happened. Or it might be possible to change the mind's algorithms so that, instead of time being linear, it is three-dimensional, like space. Consciousness could then move through the inflationary multiverse.

---

* The inflationary multiverse is a concept that follows from the idea that, shortly after the big bang, the universe started inflating exponentially like a balloon; its rate of expansion changed from place to place, giving birth to new "balloons"—new universes.

But what about the Everett multiverse, a concept distinct from the inflationary multiverse? As discussed in an earlier chapter, consciousness can indeed move through the Everett multiverse upon death. Future technology might enable us to develop the tools to control such journeys. If so, you would be able to walk through time just like you walk through space.

Either way, after creeping along for billions of years, life would finally escape from its corporeal cage.

And so we can add a ninth principle of biocentrism:

PRINCIPLES OF BIOCENTRISM

First principle of biocentrism: What we perceive as reality is a process that involves our consciousness. An external reality, if it existed, would by definition have to exist in the framework of space and time. But space and time are not independent realities but rather tools of the human and animal mind.

Second principle of biocentrism: Our external and internal perceptions are inextricably intertwined. They are different sides of the same coin and cannot be divorced from one another.

Third principle of biocentrism: The behavior of subatomic particles —indeed all particles and objects—is inextricably linked to the presence of an observer. Absent a conscious observer, they at best exist in an undetermined state of probability waves.

Fourth principle of biocentrism: Without consciousness, "matter" dwells in an undetermined state of probability. Any universe that could have preceded consciousness only existed in a probability state.

Fifth principle of biocentrism: The structure of the universe is explainable only through biocentrism because the universe is fine-tuned for life—which makes perfect sense as life creates the universe, not the other way around. The "universe" is simply the complete spatiotemporal logic of the self.

Sixth principle of biocentrism: Time does not have a real existence outside of animal sense perception. It is the process by which we perceive changes in the universe.

Seventh principle of biocentrism: Space, like time, is not an object or a thing. Space is another form of our animal understanding and does not have an independent reality. We carry space and time around with us like turtles with shells. Thus, there is no absolute self-existing matrix in which physical events occur independent of life.

Eighth principle of biocentrism: Biocentrism offers the only explanation of how the mind is unified with matter and the world by showing how modulation of ion dynamics in the brain at the quantum level allows all parts of the information system that we associate with consciousness to be simultaneously interconnected.

Ninth principle of biocentrism: There are several basic relationships—called "forces"—that the mind uses to construct reality. They have their roots in the logic of how the various components of the information system interact with each other to create the 3-D experience we call consciousness or reality. Each force describes how bits of energy interact at different levels, starting with the strong and weak forces (which govern how particles hold together or fall apart in the nucleus of atoms) and moving up to electromagnetism and then gravity (which dominates interactions on astronomical scales such as the behavior of solar systems and galaxies).

THE OBSERVER DEFINES REALITY 14

We are not only observers. We are participators.
—**John Wheeler**

P hysics is changing.

In fact, the field's biggest shift in human history may be occurring right now.

So far, our exploration of biocentrism and the evidence supporting it has mostly revolved around interpretational issues concerning quantum mechanics, whose ultimate understanding seems to require bringing consciousness into the game. In this chapter, the science behind biocentrism makes the leap from connecting-the-dots logic to newly discovered hard proofs within mainstream or consensus physics. And it makes that leap by way of the deepest nagging question in all of physics—namely, how to reconcile quantum mechanics and general relativity. To offer a bit of background, quantum mechanics works exquisitely well in describing the behavior of nature on one level, while general relativity is peerless in revealing cosmic behavior on the scale beyond quantum's grasp. Unfortunately, the two theories are fundamentally incompatible.

The problem goes far beyond "this tool only works on the small scale, while that tool operates on the large scale." After all, there'd be nothing wrong with science having a toolbox equipped with a range of useful gadgets. No, the issue is not that we need different mathematical tools to explain quantum and macroscopic systems, but rather that while these systems are ostensibly connected as part of one larger system, our cosmos, they seem to operate under two entirely different sets of rules and cannot communicate with each other.

As an example, figuring out where the moon will be tomorrow at noon requires knowing the laws of gravity, the shape of the lunar orbit, the moon's mass, and information about where it's been observed to be on past occasions. The moon's behavior proceeds according to the same laws and logic that prevails in the motion of the objects in our everyday lives, like when a friend tosses us his car keys from across the room.

But say we want to know where a particular electron will be at noon. We find classical science to be of no help. Worse, the logic of electron behavior is not the same as the logic of the visible objects around us, the moon included. Instead, we find that the electron somehow occupies numerous locations at the same time, even though it's a fundamental particle that cannot under any circumstances split itself up. To answer our question, we must use equations that reveal only the probabilities of it appearing here, here, and there, with no definite, ironclad future position ascertainable. And it remains frustrating, because even once noon arrives, what happens and where the electron appears depends on how the observer plans to observe it. With the moon, its location can be nailed down precisely by a visual sighting, a radar reflection, or even by measuring how its gravity affects passing spacecraft. With our electron, on the other hand, how we make the measurement will alter where it is.

As scientists began to study the particles and bits of energy that make up the larger structures around us, it was no small revelation that disparate science and math equipment—now called classical science and quantum mechanics—was needed, depending on what kind of object was in our crosshairs.

Reality appeared to have an irreconcilably dual nature, with general relativity seemingly the correct quantitative description of the world at large—including on the huge scales spanning the space between stars and the galaxies—and quantum mechanics describing reality at the scale of individual molecules and within atomic structure itself. For a while, this was accepted with a shrug. Quantum mechanics was new—eventually, it was assumed, we'd figure it out. Today, these two pillars of modern physics, having reached a mature age of nearly a full century, are understood in ever-increasing detail, and they have seen their theoretical predictions confirmed by countless experiments. Both theories find numerous practical applications in everyday life, such as GPS in the case of Einstein's special relativity, and transistors and microprocessors in the case of quantum mechanics.

But even after a century of continuous experimentation and gains in knowledge, we've come no closer to understanding how quantum mechanics and general relativity are compatible—how exactly physics-at-large and physics-at-small "talk" to each other.

Among the many benefits of untangling this would be the clarification of the most enigmatic of the four fundamental forces: gravity. Three of the four fundamental forces can be described by quantum mechanics; only gravity cannot. Instead, it can only be described via the classical physics of general relativity, and even that is an imperfect fit. Reconciling QT and relativity could tell us how gravity, this force with infinite reach, the force that most directly affects us in our everyday lives—whether by gluing us to our home world or periodically injuring the clumsy and unlucky through falls—might fit into the rules of quantum mechanics that appear to operate everywhere else.

In August of 2019, as this chapter was being written, a new study published in the prestigious journal *Science* showed that Einstein's theory of gravity had been proven correct (once again!). In this case, scientists used the supermassive black hole at the center of the Milky Way to test his theory of general relativity, both one of the towering achievements of the twentieth century and the currently accepted description of gravitation in modern physics.

"Einstein's right, at least for now," said Andrea Ghez, a lead author of the paper. "Our observations are consistent with Einstein's theory of general relativity. However, his theory is definitely showing vulnerability. It cannot fully explain gravity inside a black hole, and at some point we will need to move beyond Einstein's theory to a more comprehensive theory of gravity."

There is a whole field of physics dedicated to attempting to explain gravity by way of quantum mechanics—what's known as *quantum gravity*. At the core of the incompatibility between QT and relativity, the two foundation posts of modern theoretical physics, is the "non-renormalizability" of quantum gravity. And, in a stunning turn of events, it turns out that to address this issue, one needs to incorporate something that modern theoretical physicists working in this area have largely ignored until now.

You guessed it: observers.

"Non-renormalizability" is a term from the jargon of cutting-edge physics, and the concept is complex, but it still boils down to physics and math at one scale not working in any way at a different scale. A non-renormalizable theory is one where the particular phenomenon or group of phenomena it describes remains mathematically well under control at one particular spatial scale (say, a small one), while at a different scale (say, a larger one) this control may be completely lost—meaning the math and physics no longer work. A non-renormalizable theory is somewhat akin to a magnifying glass. Imagine a naturalist using such a glass to study an object: At the right distance, this powerful tool lets the naturalist see the object more clearly. When the magnifying glass is moved away from the object, however, the image is slightly distorted. Shift the magnifying glass even further, and the image becomes totally unrecognizable. Similarly, we do not really know what the correct structure of reality is when described by a non-renormalizable theory: according to such a theory, this structure changes drastically when we proceed from studying reality at one scale to probing it at another. Indeed, the language—that is, the physics and mathematics—we need to explain what we see gets increasingly complicated, eventually becoming *infinitely, uncontrollably complicated* at some sufficiently large scale.

The behavior of the gravitational force is explained very well by relativity, but the smooth continuum of relativity's spacetime and the chunky, quanta-based world of QT don't play well together. When we try to use the language of quantum mechanics to describe gravity, everything we can measure as observers (for example, the curvature of spacetime or the energy stored in a unit volume of matter) starts to blow up infinitely and uncontrollably, and we quickly get lost in mathematical infinities without any possibility of making meaningful predictions or defining measurable quantities.

The frustration of physicists encountering this insoluble situation over the past century might be better appreciated if we imagined what it might be like if everyday objects behaved the same way. When the Scottish genius John Dunlop invented bicycle tires late in the nineteenth century, he was very familiar with the properties of rubber. But imagine if his tires worked as planned only as long as the bike traveled under five miles per hour. What if, the moment a rider exceeded that leisurely speed, the rubber turned rigid instead of flexible, and it abruptly became so sticky it glued the wheel to the road? And what if no scientific investigation could explain this dramatic shift from functional to worthless when seemingly irrelevant conditions changed? Imagine poor John's bewilderment! Well, the way quantum gravity theories transform into uselessness, but only at certain scales, has perplexed the greatest theoretical physicists to the same degree.

Now, however, new research published in the *Journal of Cosmology and Astroparticle Physics* (JCAP)* by theoretical physicist Dmitriy Podolskiy, in collaboration with one of the authors (Lanza), and Andrei Barvinsky (one of the world's leading theorists in quantum gravity and quantum cosmology) has revealed something remarkable. Namely, that this exasperating incompatibility between quantum mechanics and general relativity vanishes if one takes the properties of observers—us—into account.

---

* JCAP is one of the leading journals in cosmology and astrophysics—which published Stephen Hawking's groundbreaking work on the evolution of the early universe. Please see Appendix 3.

In classical physics, it's assumed that we are able to measure the physical state of an object of interest without perturbing it in any way. This sounds reasonable if we follow our everyday intuition. When we look at an airplane to determine its position with respect to the ground (Did it already take off? Is it landing?), we have zero influence on its state unless we ourselves are the pilots or flight controllers. If the states of physical objects are unperturbed by our measurements, then probing them or their responses to some external influence allows us to accurately create a physical theory describing them precisely.

But in the realm of quantum, as we've seen throughout this book, things are considerably more complicated—properties are a matter of probability, and our measurements and observations not only perturb reality, but create it. Quantum gravity is no exception. Our friend Wheeler coined the term "quantum foam" (sometimes also called "spacetime foam") to refer to what spacetime might be like at the quantum level, full of tiny fluctuations rather than exhibiting the seeming smoothness we observe at larger scales. These fluctuations cause tiny alterations in the paths of particles, and by looking for these, scientists can measure this quantum gravitational spacetime. If many observers continuously measure the state of this wobbling quantum gravitational spacetime foam (in particular, to determine how much the spacetime is curved) and then exchange information about the outcomes of their measurements, it turns out that the presence of the observers themselves significantly perturbs the structure of physical states of matter and spacetime itself. In grossly simplified language: it matters enormously to the perceived laws of reality how many of us are here studying or probing it, and what we communicate to each other about the results of our measurements.

The nature of this unusual phenomenon goes back to an important discovery made in the late 1970s by Italian physicist Giorgio Parisi and his Greek collaborator, Nicolas Sourlas. The rough technical language used by the authors states that a physical system existing in (D + 2) spacetime dimensions in the presence of a disorder influencing its physical states is largely equivalent to a similar system living in D spacetime dimensions without any disorder. To put this more simply, when disorder/random components are added to a physical system, its complexity will increase.[†] But what does this really mean, and what does it tell us?

First, let's clarify "disorder." When talking about the presence of a disorder, Parisi and Sourlas meant applying a random external force to the physical system of interest at different points of spacetime. A case of such "disorder" occurs when a number of observers simply measure the state of the physical system under consideration (for example, momentum, energy density, or, when the system is spacetime itself, curvature of this spacetime) at random points.

Second, recall that dimensionality of an object or a space is the number of completely independent directions we can move along the object or in space. For example, a very narrow wire is basically a one-dimensional object, since it essentially offers only one direction to wander—along its length. A sheet of paper is two-dimensional (it has both length and width), while a cube or a cylinder is three-dimensional (they are characterized by their length, width, and height). As Einstein taught us, the spacetime we are living in is four-dimensional, the role of the fourth dimension being played by time.

We can now more clearly formulate the conclusion of Parisi and Sourlas: Generally, any presence of observers distributed across spacetime and randomly measuring the state of reality leads to an *effective increase in dimensionality* of the spacetime in which the physical system of interest resides.

Okay, but what does this have to do with the "non-renormalizability" of gravity and the efforts to unify the two pillars of physics?

Well, as it turns out, "non-renormalizability" and "dimensionality of spacetime" are intimately related. Typically, the higher the dimensionality of spacetime a theory relies upon, the more likely it is that this theory is non-renormalizable.

For example, consider "quantum electrodynamics," which investigates the quantum dynamics of electromagnetic fields and their

---

[†] Surprisingly, as we learn in physics on so many examples, low-dimensional systems are almost always more complicated than higher-dimensional ones, having in particular much more complicated dynamics.

interaction with electric charges. The theory of quantum electrodynamics, which covers 95 percent of all physical phenomena we see around us, was developed by Richard Feynman and other physicists back in the 1950s, and it happens to remain perfectly under control at all spatial scales (meaning, it's renormalizable) as long as the dimensionality of spacetime is two, three, or four. It ceases to be well behaved (becomes non-renormalizable) if the number of spacetime dimensions is five or higher.[‡] Similarly, the Standard Model of high-energy physics—which includes the weak, strong, and electromagnetic interactions that surround us in everyday life—breaks down whenever the number of dimensions gets higher than four.

Physicists invented a special term for this threshold: *the upper critical dimension*. A theory becomes non-renormalizable (that is, it breaks down/ the math won't work consistently) if the dimensionality of the spacetime it is defined within is higher than this upper critical dimension. For the majority of physical interactions (the weak, strong, and electromagnetic), this upper critical dimension turns out to be four—exactly coinciding with the dimensionality of the spacetime we are actually living in! Ultimately, this is why theoretical physics has been so successful in describing numerous physical phenomena occurring in the quantum world of high-energy physics.

However, our luck runs out with quantum gravity. The critical number of spacetime dimensions above which theories of quantum gravity start behaving uncontrollably badly is two—one dimension for time and another one for space. Since the dimensionality of spacetime we are living in is four, this means that quantum gravity is two spacetime dimensions away from being a controllable theory.

---

‡ A side note: How can the dimensionality of spacetime be higher than four? One needs to be a bit of an abstract thinker to imagine this. Consider again a flat sheet of paper (a two-dimensional object), which is put somewhere in a three-dimensional space (placed on your table). An ant moving along the paper would not know that the real world around him is three-dimensional (actually four-dimensional, if we take time into account). The same logic might be applied to us: our four-dimensional world can probably be embedded into a five-dimensional one, and we would never know the difference.

Now, if we follow the logic of Parisi and Sourlas outlined above—where a system in spacetime with dimensionality of D + 2 *with* disorder present roughly translates to a system in spacetime with dimensionality D *without* disorder—we see that quantum gravity in four spacetime dimensions, in the presence of a large number of observers (disorder), is in fact the same as quantum gravity in a spacetime with two fewer dimensions. In other words, just two. We have perfect control over such a theory, know very well how it works at all scales, and hence this resolves the long-standing paradox of incompatibility between general relativity and quantum mechanics.

Let's now examine the fascinating consequences of this revelation to see the hard-science proof underlying the details of how the presence of observers not merely influences but *defines* physical reality itself.

First of all, if one believes that the reality described by a combination of Einstein's theory of general relativity (working at large spatiotemporal scales) and quantum mechanics (working at small scales) exists and makes nature operate smoothly, then said reality must also contain observers in one form or another. Without a network of observers continuously measuring the properties of spacetime, the combination of general relativity and quantum mechanics stops working altogether. So it is actually *inherent to the structure of reality* that observers living in a quantum gravitational universe share information about the results of their measurements and create a globally agreed-upon cognitive model of it.

Remember, once you measure something (for example, the location of an electron in a particle physics experiment, the length of an electromagnetic wave, or the curvature of spacetime defining the gravitational pull between two bodies), the probability wave of measuring the same value for the already-probed physical quantity becomes "localized" or simply "collapses" (see figure 14.1). This means that if you keep measuring the same quantity over and over again, keeping in mind the result of the very first measurement, you will continue to see a rather similar outcome of the measurement process. A much-simplified illustration of this is a famous thought experiment by Richard Feynman: Consider a wall with two slits and two detectors of electrons (such as photo plates) behind them. When we continuously send electrons toward the wall, both photo plates eventually have imprints of electrons hitting them; once an electron hits the photo plate, the imprint on it remains forever, and we shall keep seeing this imprint at second and subsequent looks at the photo plate. Physicists say that the electron's one-particle wave function "collapses" at the moment the electron hits the photo plate—or, in other words, that the process of "decoherence" happens then. While this outcome seems pretty deterministic in contrast to the probabilistic way we know quantum mechanics operates, its quantum nature will in fact be reflected in a wavy interference pattern produced on the plates by multiple electrons hitting them one after another.

Without any measurements, the waves of probability for various observable quantities (such as spacetime curvature) to possess particular, fixed values will be blurred, colliding and scattering off each other so that physical reality remains a wobbly, undetermined mess—the underlying quantum foam. Measurement or a sequence of measurements collapses these waves of probability and distills them out of the quantum blur.

If you learn from somebody the outcomes of their measurements of a physical quantity, knowing these outcomes will also influence the outcomes of your own measurements, freezing the reality according to a consensus between your measurements and those of other observers. In this sense, a consensus of different opinions regarding the structure of reality defines its very form.

Recall that time itself, as well as the direction of the arrow of time, becomes defined due to the process of wave function collapse (or decoherence). Once such temporal collapse happens, one can start asking questions about the *dynamics* of the process of decoherence for other physical quantities that we as observers can measure. These dynamics—how quickly the collapse of quantum blur toward a particular realization of measurable quantities happens, how long it stays collapsed, the detailed structure of the probability waves defining observed reality—strongly depend on how the measurements or observations by different observers are distributed within spacetime. If there are many observers and the number of observations made by them is very large, the probability waves of the measurement of a macroscopic quantity remain largely "localized," not spreading much, and reality is largely fixed, deviating slightly from the consensus only every once in a while. (A rough quantitative criterion for this is that the characteristic spatiotemporal scale of an object or a process you are studying should be larger than a characteristic interval between measurement events; for example, if measuring the gravitational pull of our planet, we should make measurements at intervals shorter than the time needed to traverse the diameter of Earth at light speed, which is also gravity's speed.) From one location to another within the background spacetime, both the speed with which the probabilistic structure of the universe collapses toward the consensus and the possibility of deviations from that consensus vary slightly depending upon how densely packed the observation events are, how many observers are present, how quickly they share information about their measurements with each other, and how strongly they interact with parts of the objective reality that they are trying to measure (figure 14.1). This variation is testable. It can be checked both by performing real experiments and numerical simulations for various quantum-mechanical systems—it has been checked numerically already and will be checked experimentally in the near future.

The numerical simulations that served as tools for assessing the physics of these phenomena were the "Monte Carlo methods," which are often used in physical and mathematical problems, especially when it is difficult or impossible to use other experimental approaches. This simulation method was first outlined and used successfully during the Manhattan Project in the development of nuclear weapons, for example as a means of investigating how neutrons travel through radiation shielding. In the case of today's high-intensity physics-related problems, Monte Carlo methods allow us to simulate systems with many coupled degrees of freedom, such as fluids, disordered materials, strongly coupled solids, and cellular structures, the only drawback being the huge amount of required computer power. The simulations used in the new Podolskiy-Barvinsky-Lanza study were performed using the massive MIT computer cluster.

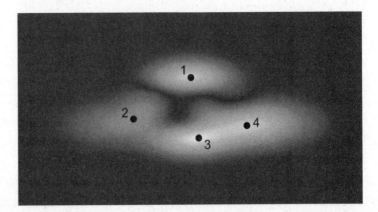

Fig. 14.1 Consensus defining reality—the probability of measuring a given value of spacetime curvature for four observers located close to each other. Observers 1 and 2 do not know about each other and are probably separated by large distances; as a consequence, the results of their measurements are slightly different. Observers 3 and 4 share information about their measurements (perhaps these two points even describe the same observer!) and the probability of measuring the same spacetime curvature measured by one of them will likely be the same for the other.

You might wonder what would happen if there were only one observer present in the whole universe. How does it change the physical picture described above? Do the probabilistic waves describing the physical reality of our universe collapse in that case—and does quantum gravity become a workable theory? The answer depends on whether the observer is conscious, whether she has memory about the results of probing the structure of objective reality, and whether she builds a cognitive model of this reality.

For a conscious observer, the sequence of measurements she performs is akin to a random network of measurement events with information describing the outcomes of these measurements being passed between the events—the world line of a single observer is nothing but a sequence of points/events that are very close to each other in spacetime. In other words, a single conscious observer can completely define this structure, leading to a collapse of the waves of probability, describing it as a particular realization of the quantum blur largely localized in the vicinity of the cognitive model the observer builds in her mind throughout her life span. As experimental results confirm this, we will be reshaping our understanding of reality in a way that is long overdue—seeing how intimately we are connected with the structures of the universe on every level.

With this new Podolskiy-Barvinsky-Lanza study, solid evidence seems finally to have arrived showing that *observers ultimately define the structure of physical reality itself.*

Most significantly, this study relies and builds upon the existing, cutting-edge scientific theories accepted by almost all physicists. Yet these accepted physical theories of the universe that embrace everything from Einstein to Hawking to string theory are based on something being "out there" beyond ourselves, whether fields, quantum foam, zooming photons, or whatever.

The conclusion—to which not only this entire book, but indeed the long history of physics itself, has been inexorably leading—is that the world is observer defined *regardless* of whether one believes in multiple universes or simple wave function collapse, whether one embraces Copenhagen, is attracted or repelled by string theory, and all the rest. There is really no getting around the fact that the cosmos is biocentric.

We are living through a profound shift in worldview, from the long-held conception of the physical world as a preformed entity that just exists, fully formed, "out there," to one in which it belongs to the living observer. To us.

And so we can add a tenth and eleventh principle of biocentrism:

PRINCIPLES OF BIOCENTRISM

First principle of biocentrism: What we perceive as reality is a process that involves our consciousness. An external reality, if it existed, would by definition have to exist in the framework of space and time. But space and time are not independent realities but rather tools of the human and animal mind.

Second principle of biocentrism: Our external and internal perceptions are inextricably intertwined. They are different sides of the same coin and cannot be divorced from one another.

Third principle of biocentrism: The behavior of subatomic particles —indeed all particles and objects—is inextricably linked to the presence of an observer. Absent a conscious observer, they at best exist in an undetermined state of probability waves.

Fourth principle of biocentrism: Without consciousness, "matter" dwells in an undetermined state of probability. Any universe that could have preceded consciousness only existed in a probability state.

Fifth principle of biocentrism: The structure of the universe is explainable only through biocentrism because the universe is fine-tuned for life—which makes perfect sense as life creates the universe, not the other way around. The "universe" is simply the complete spatiotemporal logic of the self.

Sixth principle of biocentrism: Time does not have a real existence outside of animal sense perception. It is the process by which we perceive changes in the universe.

Seventh principle of biocentrism: Space, like time, is not an object or a thing. Space is another form of our animal understanding and does not have an independent reality. We carry space and time around with us like turtles with shells. Thus, there is no absolute self-existing matrix in which physical events occur independent of life.

Eighth principle of biocentrism: Biocentrism offers the only explanation of how the mind is unified with matter and the world by showing how modulation of ion dynamics in the brain at the quantum level allows all parts of the information system that we associate with consciousness to be simultaneously interconnected.

Ninth principle of biocentrism: There are several basic relationships—called "forces"—that the mind uses to construct reality. They have their roots in the logic of how the various components of the information system interact with each other to create the 3-D experience we call consciousness or reality. Each force describes how bits of energy interact at different levels, starting with the strong and weak forces (which govern how particles hold together or fall apart in the nucleus of atoms) and moving up to electromagnetism and then gravity (which dominates interactions on astronomical scales such as the behavior of solar systems and galaxies).

Tenth principle of biocentrism: The two pillars of physics—quantum mechanics and general relativity—can only be reconciled by taking observers, us, into account.

Eleventh principle of biocentrism: Observers ultimately define the structure of physical reality—of states of matter and spacetime—even if there is a "real world out there" beyond us, whether one of fields, quantum foam, or some other entity.

DREAMS AND MULTIDIMENSIONAL REALITY 15

It was just a dream, but so real that life could learn from it.
—**Matej Bor**

N
ow that we're nearly at the end of our story, and we've seen the hard evidence for biocentrism, let's take a break from concepts like "renormalizability" that would bring conversation to a screeching halt at any social gathering and instead look at what it all means for a familiar, everyday (or every night) phenomenon, one that nonetheless has some intriguing implications for our study of awareness: dreams.

The secrets dreams can help unlock ultimately derive from the basic and obvious fact underlined by biocentrism—that reality is always a process that involves our consciousness.

We assume the everyday world is "out there" in a more real or independent sense than is the world of our dreams, that we play a lesser role in its appearance. Yet experiments show that day-to-day reality is every bit as observer dependent as dreams are.

As we've said over and over throughout this book, everything we experience is simply a whirl of information occurring in our heads. And by "everything," we mean that literally and absolutely.

This leaves no room for external frameworks. Indeed, as we've seen, biocentrism tells us that space and time are not actual entities, but rather terms that designate the tools our mind uses to assemble information. They are among the most critical keys to consciousness, and they explain why, in experiments with particles and the properties of matter itself, they are always relative to the observer, as opposed to being objective, stand-alone absolutes.

As we go about our lives, we take for granted the way our minds put everything together because the process is effortless, and its underlying mechanisms are baked in, hidden, and automatic. But you might not have suspected that this same process of fashioning a seemingly external 3-D reality is the one underlying dreams. Since the realms of dreams and wakeful perception are usually classified separately—with only one of them regarded as "real"—they're rarely part of the same discussion. But there are interesting commonalities that give us clues as to how our consciousness operates. Whether awake or dreaming, we are experiencing the same process even if it produces qualitatively different realities. During both dreams and waking hours, our minds collapse probability waves to generate a physical reality that comes complete with a functioning body. The result of this magnificent orchestration is our never-ending ability to experience sensations in a four-dimensional world.

This genesis of the dream realm begins with the simple fact that all organisms sleep. We cannot live our wakeful lives without also sometimes sleeping; experiments show that an organism deprived of sleep will die. Sleep consists of periods of dreams, called REM sleep,* and also of periods without dreams, non-REM. When we awaken we often remember our dreams, but we have no memories of whatever was going on during the non-REM period of sleep. That is because, during a

---

* REM stands for "rapid eye movement."

non-REM period, the wave packet is so widely spread that most of its branches are decoupled from each other, with no interactions or entanglement between them. On awakening, you find yourself in one of those branches, experiencing your familiar world. During dreams, however, the branches of the spreading wave function are not totally independent and decoupled; once back in consensus reality, then, memory has access to those other branches/worlds.

We've all had the experience of waking up from a dream that seemed every bit as real as everyday life, even if the sights and experiences were ones entirely unfamiliar to our waking selves. "I remember," recalled one of the authors (Lanza) in a *Huffington Post* article, "looking out over a crowded port with people in the foreground. Further out, there were ships engaged in battle. And still further out to sea was a battleship with radar antenna going around. My mind had somehow created this spatiotemporal experience out of electrochemical information. I could even feel the pebbles under my feet, merging this 3-D world with my 'inner' sensations. Life as we know it is defined by this spatial-temporal logic, which traps us in the universe with which we're familiar. Like my dream, the experimental results of quantum theory confirm that the properties of particles in the 'real' world are also observer determined."

We dismiss dreams because they end when we wake up, and also because they are largely enigmatic. Dream labs and researchers engaged in investigations over decades still cannot explain why dreams during the night's first few hours revolve around the recent day's events, while later dreams are far more surreal in content. Specialists still don't fully understand why we dream for only about two hours total, or why the emotions experienced during dreams are overwhelmingly negative. Or why the five-minute dreams typical of 11 PM eventually morph into protracted predawn reveries, lasting ten times longer.

Nonetheless, the transient duration of the experience is poor reason to diminish it. Certainly we don't think our experience of day-to-day life is less real because it ends when we fall asleep or die. It's true we don't remember events in our dreams as well as we do those occurring in waking hours, but the fact that Alzheimer's patients may have little memory of events doesn't mean their experience is any less real. Or that individuals who take psychedelic drugs don't experience physical reality during their "trips," even if the spatiotemporal events they experience are distorted or they don't remember all of the events when the drugs wear off.

We might also dismiss dreams as unreal because dream researchers have found them to be closely associated with specific patterns of brain activity. But are our waking hours unreal because they're similarly linked with neural activity in our brain? Certainly, the biophysical logic of consciousness—whether during a dream or waking hours—can always be traced backward to something, whether in space to neurons or in time to the big bang.

Dreams must be far more than the spontaneous, random firing of neurons that some insist they are. They must likewise be far more than the activation of random memories already contained in the brain's neurocircuitry. True, dreams often contain a mix of emotions and things we have previously experienced, but as we've already pointed out, in dreams there are often people, faces, and interactions that the dreamer has never experienced before. A dream is an instantaneous, nonstop narrative that often seems as real as real life itself. How could this tapestry of enormously complex interactions and scenarios be the result of nothing but random electrical discharges? In dreams we are not just watching an "external world" and passively imprinting memories in our neural circuitry. How is it possible for the brain to do this? How are all the components of the experience fabricated from scratch? While dreaming, we are not observing events and perceiving stimuli. We are in bed, asleep—yet our minds are able to flawlessly create new people and settings and have them all interact effortlessly in four dimensions. We are witnessing an awesome occurrence: the ability of the mind to turn pure information into a dynamic multidimensional reality. You are actually creating space and time, not just operating within it like a character in a video game.

While it is easier to appreciate the astounding nature of this process when it comes to dreams, it is the same process described throughout the book as applying to our nondream lives. According to biocentrism, we are *always* not just observing but creating reality.

And, as in "real" life, in dreams the collapse of probability waves is a critical component of the multidimensional realities the mind creates.

We collapse probability waves in our dreams just as we do when we are awake. During dreams, however, the brain has fewer limitations since it needn't obey sensory inputs that themselves are limited by physical laws, and thus the mind can generate experiences unlike the consensus world we're aware of during the day.

In Chapter 14 we discussed how the presence of extended networks of observers defines the structure of physical reality itself. In dreams, we leave the consensus universe and can experience an alternate cognitive model of reality—very, very different from the one shared by other observers while awake. In dreams, the fine structure of the wave function of the universe around us is delocalized and thus largely unstable. This explains why you often have more power while dreaming; the values of observables representing the basis of reality are more fluid. As also discussed in Chapter 14, the presence or absence of a network of observers influences the very dimensionality of the universe. In dreams, the number of dimensions can also change, depending on the specific information recruited into the mind's construction.

Dreams are often very vivid, but Lanza recalls one, in particular, that stands out from all his others. The resolution of this dream was unmatched by anything he had experienced before—it was like going from watching a grainy old movie to watching one in ultra-high definition. In the dream, he experienced an extra spatial dimension that allowed him to see (with crystalline clarity) both the inside and outside of the objects he observed from all sides/directions *at the exact same time.* For about two or three minutes after he woke up, before the dream had completely faded from his mind, he was able to go back and forth between experiencing the four-dimensional construction (one temporal + three spatial dimensions) of his waking reality and the five-dimensional construction (one temporal + four spatial dimensions) of his dream. Although he can still remember some of the memories from that transition period, this 5-D world cannot be experienced in the four-dimensional consensus reality that you and I (and everyone reading this book) are collectively a part of.

Biocentrism says space and time are tools of the mind, and dreams seem to be only further proof of the truth of this statement. For, if space

Fig. 15.1 What is it like to experience a 5-D (one temporal + four spatial dimensions) reality? Dreams demonstrate the capacity of the mind to construct multidimensional realities, both 4-D (one temporal + three spatial dimensions) and, in some cases, even 5-D (one temporal + four spatial dimensions). One manifestation of the latter is the ability to see the interior and exterior of an object from all spatial perspectives simultaneously at each instant in time.

and time were really external and physical as is popularly believed, then how would it be possible to create something absolutely indistinguishable from them within the confines of one's dreaming brain?

In previous chapters, we've explained that what we perceive as reality is a result of wave function collapse. Wave function is a mathematical description of conscious experience associated with physical measurements and observations of the world. We collapse the wave function during observation, using our senses of sight, hearing, touch, and so on. In an awakened state we perform observations very frequently, almost continuously, and thus repeatedly collapse wave function, which would otherwise (i.e., in the absence of observations) begin to spread in the abstract "Hilbert space"[†] of possibilities. A simple model of this is the textbook example of the spreading of a wave packet described in Chapter 10. When we go to sleep, our observations or measurements cease, and the wave function starts spreading so that it incorporates many possible "worlds" or experiences. We then have the potential to "create" any of those possible worlds by collapsing the wave function accordingly. In sleep, we are wandering in Hilbert space and experiencing wave function collapse in numerous diverse ways. Eventually, the wave function of our experiences collapses so that we find ourselves awakened within the same bed and the same room that we remember going to sleep in the previous evening. We recall who we are, our name, and our memories of previous life events. We think that our experience at night was only a dream, and that it wasn't real. But, as explained above, dreams and what we perceive as reality are basically of the same nature. And such a view has its support in quantum mechanics.

Thus, I wake up in the morning as this person, living in this house, town, country. But my wakening was just one possible collapse of the grand wave function into a definite world of my experience, as discussed in Chapter 7. There are many other possible ways the grand wave function can collapse. It can collapse so that it becomes a wave function describing experiences of person A, or it can collapse so as to describe the experiences of person B, or anybody else, including an animal, a bird, fish, or any living creature.

This has nothing to do with multiple, separate consciousnesses existing in the same world. In each case of the wave function collapse, there is a different world with a unique, single consciousness. In one such world, consciousness experiences the life of person A while all other persons are perceived as being "external" to A, as described in Chapter 7. In another world, the same consciousness experiences the life of person B, while all other people and animals are, like trees, houses, and other inanimate things, perceived as being "external" to B.

[†] In mathematics and physics, the concept of "space" can have an abstract meaning that goes far beyond our usual notion of space. Quantum mechanics operates with the concept of Hilbert space, which is the space of all possible wave functions, including the collapsed ones corresponding to definite experiences.

"What Is It Like to Be a Bat?" is the title of an article by Thomas Nagel published in 1974 in the *Philosophical Review*. Nagel wrote, "The fact that an organism has conscious experience *at all* means, basically, that there exists something which is: what it is like to *be* that organism. So fundamentally an organism has conscious mental states if and only if there is something that it is like to be that organism."

Throughout this book we are adopting the thesis that this "something" is wave function interpreted as a mathematical description of consciousness. Collapsed wave function describing the experiences of person A corresponds to Nagel's "something it is like to be [person A]."

In your awakened state, you experience your consensus reality. Then you go to bed, fall asleep, and start dreaming. And when you wake up, you find yourself again existing as a person in a consensus reality. Through dreams you enter alternate worlds and switch from one consensus reality to another, from experiencing the life of one organism to that of another. Once awake, you can find yourself as being any person, at any time, without having memories about ever being another person or animal. You can even find yourself as a newborn, without any ideas about the reality you are living. If so, gradually, piece by piece, you discover your reality, your world. By observing your world, you keep collapsing probability waves, and thus you effortlessly create an ever-more detailed world that includes comprehensive reinforcing memories. The observations also include what others tell you about the world and its history, and so you build your consensus reality.

It is amazing how far we've come by following the implications of quantum mechanics in an unbiased way. By adopting the idea that wave function is a mathematical description of experience, we arrive at the unification of everyday reality and dreams. And our dreams provide further vivid support for what we've said about the ongoing wave collapse that manifests itself as unending conscious experience. Persistent puzzles regarding quantum mechanics, many worlds, and wave function collapse, about consciousness, reality, and our own lives and deaths, all fade away.

OVERTHROW OF THE PHYSIOCENTRIC WORLDVIEW

16

All false art, all vain wisdom, lasts its time but finally destroys itself.
—Immanuel Kant

I t has been quite a journey, for you the reader, as well as for humans in general, who have slogged through the centuries trying to figure out the universe.

We humans began with simple superstition, responding to the fact that existence seemed precious but fragile, and that life's everyday pleasures could be abruptly snatched away by a flood or some sudden disease.

So we naturally begged first the gods, and later one God, to be merciful and spare us, and this policy of whining, sniveling, and bargaining with the invisible ruling superpowers that presumably surrounded us pretty much constituted our collective worldview. Millennia elapsed, and in time the ancient Greeks and then the brilliant minds of the

Renaissance began to perceive that the world was ruled by more than supernatural whim; nature marched in rational ways, according to laws our minds could decipher.

This changed everything, and our knowledge now expanded at tremendous speed and with awe-inspiring consequences. It was no minor accomplishment when Johannes Kepler showed that the earth, moon, and planets all moved in elliptical orbits, and their future positions were not just fathomable, but predictable to a high degree of precision. We could even foresee when eclipses would darken the land. A grand order was now perceived to rule nature, and it was marvelous.

But a powerful dichotomy remained. First, the divide was placed between the heavens and us mortals on Earth; later, between us and nature. In the seventeenth century, René Descartes insisted that mind and matter were fundamentally different from each other, meaning consciousness or perception stood apart from the rest of nature. This separation of ourselves from the bulk of the universe drew favorable nods from science and the clergy alike. If we were to study the cosmos, it made sense that our own fallible perceptions should be removed from the process. And religion, of course, approved of a view that we humans were something more than mere matter.

As the universe got bigger, our own place in it grew correspondingly smaller.

As scientists struggled to pry the masses away from religion and superstition, they were happy to promulgate a worldview in which science could provide answers, and pure objectivity was possible; in other words, one where we observers were simply not very important. And when Edwin Hubble showed in 1930 that the universe consists of billions of galaxies, each containing billions of stars like the sun, with planets as common as snowflakes in a blizzard, our new collective mindset became: *How tiny we each are! How inconsequential!*

Thus, as the early decades of the twentieth century unfolded, small was "in." Insignificance had become fashionable. We individual observers now viewed ourselves as less than unnecessary. We could all vanish, and the cosmos would continue unchanged.

And doesn't most everyone you know still embrace this view?

This is why the strange experimental results seen by the originators of quantum theory were so thoroughly disquieting. Because, over and over, they showed that physical parameters like an object's position and motion *depended on the observer.*

There had of course been hints for centuries that observers might have some role to play in reality. In fact, in his *Opticks*, Isaac Newton insisted that brightness and hue are not inherent, but that each observer actually creates all the colors of the visual realm within his mind. "[T]he Rays to speak properly are not coloured," he wrote. Other scientists eventually showed that Newton was right. By the early twentieth century, physicists had established that light consists of the alternating pulses of magnetic and electrical fields. Since neither magnetism nor electricity are visible to humans, to our eyes a verdant forest canopy must be inherently blank. The fact that we see it as emerald green means that somewhere in the vast magical neurocircuitry of our brains, a "green" sensation arises, and then, by some equally marvelous mental occurrence, we "place" it out in front of our noses, in what we regard as the "external world."

Thus many scientists were indeed increasingly aware that the distinction between internal and external was artificial, and that everything perceived—whether a traffic light or an itch—arises strictly in the mind. Mind, or perception, or consciousness, or awareness, is neither internal nor external. Rather, it encompasses everything—all experience.

By the 1920s, however, many of quantum theory's originators had been dumbstruck by the discovery that the role of the observer went far beyond mere perception. There was growing evidence that it wasn't just the *visual* cosmos that was observer dependent; it turned out, the act of observation is what causes small physical objects to behave as they do, and even to pop into existence in the first place. Physicists were gaining a sudden new appreciation of the role of consciousness in how nature operates on the smallest scales.

Still, in many scientific circles, these revelations simply didn't fly, mainly because it all seemed mighty close to philosophy or metaphysics. The comparison wasn't unwarranted; the new quantum conceptions of the observer and consciousness actually paralleled many of the ancient tenets of the East. Some quantum theorists, like Erwin Schrödinger, went even further along that path by wondering where one person's consciousness ends and another's begins. Remember: "Consciousness is a singular of which the plural is unknown," he said. Mainstream science was aware that traversing this route might throw an unwelcome monkey wrench into the quasi-official, curriculum-sanctioned worldview that still effectively embraced the firm Cartesian separation between mind and matter, nature and us conscious observers.

But the tide could only be held back temporarily. In experiment after experiment, such as the famous double-slit demonstration, the "delayed-choice" setup, and countless others, the observer's importance kept manifesting itself. The results were baffling, but, after decades of ever-increasing corroboration, undeniable. This is why the famed Princeton physicist John Wheeler was able to so confidently state, "No phenomenon is a real phenomenon until it is an observed phenomenon."

And that brings us up to the era of our own lives. As we've seen, we did not arrive here out of the blue. The chapters in this volume faithfully recount the unfurling of knowledge that propelled us forward, tracing the physics story line from the genius of Isaac Newton through the major reassessments of the eighteenth and nineteenth centuries, when scientists began to discover unexpected fundamental unities in all sorts of crevices throughout the cosmos. As the march of science continued, "what we know for sure" was turned on its head again and again, with the space-and-time and matter-and-energy relationships revealed by Albert Einstein, then the still greater upheavals ushered in by the geniuses of quantum theory.

And all of this has led to the logical next step: biocentrism. Biocentrism identifies life and consciousness as the central reality of existence not out of any petty desire or dogma-driven need to elevate our own status as living beings, but because centuries of hard-won scientific knowledge and experimental data show it to be the only consistent explanation for what we see around us.

Alas, in keeping with human nature, mainstream science continues to resist a whole-scale change to its long-held worldview—one in which observers enjoy pretty much the same status as laboratory mice—even as physicists acknowledge the truth of quantum theory and turn up stranger and stranger examples of phenomena, like entanglement, that affirm its predictions. To many in the scientific community, even today, the very word "consciousness" sends up red flags, as if all observer-linked experimental results somehow invoke the supernatural, or are fringe science akin to the renegade psychedelic investigations of the 1960s.

At the same time, an ever-more educated global population turns increasingly to science for answers to the timeless mysteries that plague us. Is reality real? Are we conscious beings reducible to our physical brains? Is there life after death? Why does the universe work the way it does? What is my place in it? Mainstream science has had little success in tackling such questions. But the biocentric paradigm indeed provides answers. What was needed to move the corpus of the scientific establishment, and change the public consensus once and for all, was hard evidence in support of biocentrism's conclusions.

In service of these conclusions, the first two biocentrism books invoked logic, philosophical arguments by great thinkers of both ancient and modern times, and detailed accounts of scientific experiments. The present book cements this by offering more detailed explanations of the science behind the theory, and published papers pointing toward its truth.

Much indirect or secondary evidence has long supported a biocentric view of the cosmos. For instance, it is hard to escape the fact that some two hundred basic physical parameters that are unvarying throughout the universe, such as the strength of the electromagnetic force *alpha*, all have precisely the values necessary to allow life to exist. Sure, there's nothing to say that this *couldn't* be pure coincidence. But in science, researchers are justifiably fond of invoking "Occam's razor," which is the principle that the simplest explanation is usually the correct one. So while it *could* perhaps be just an accident—and an "accident" (or its synonym, "random occurrence") is indeed how mainstream science continues to explain it!—that all two hundred of these physical constants are aligned perfectly for stars to shine, multiple kinds of atoms to exist, and life to arise, blithely accepting such an unlikely concurrence would leave a lot of ugly stubble on science's chin. But if you accept instead the biocentric theory, that life is central, then no other values for these physical constants could ever have been possible, end of story—and what could be simpler and more Occam approved?

Alas, how can science set up an experiment where one physical system is placed in the presence of consciousness while another is kept separate from any observer awareness so as to perform the kind of standard A/B comparison needed to see how observation actually affects things?

Happily, the double-slit experiment and innumerable variations, repeated thousands of times for decades, have already given us just this comparison. Time and again, results consistently show that an observer's

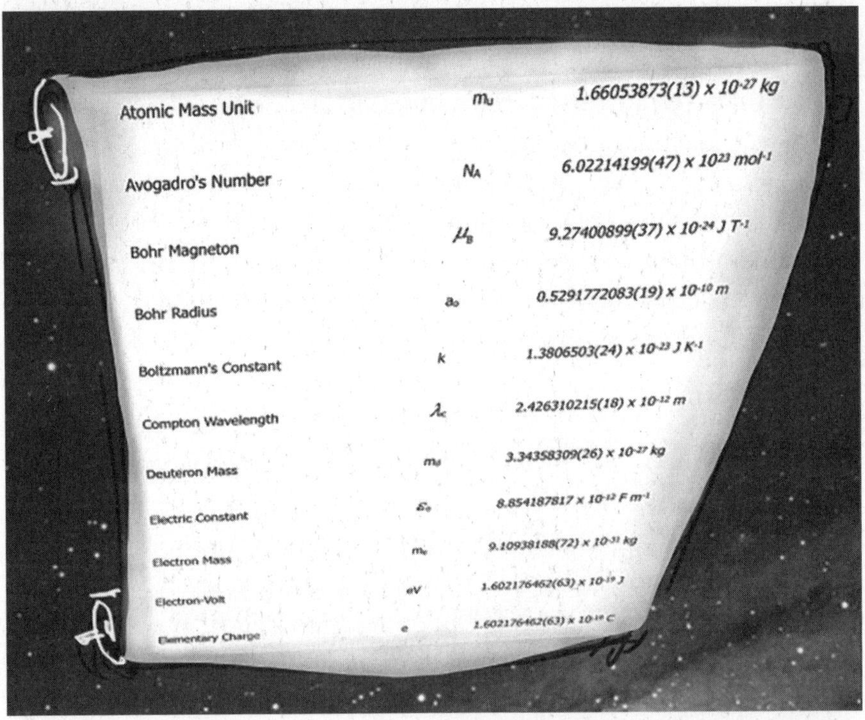

Atomic Mass Unit	m_u	$1.66053873(13) \times 10^{-27} \, kg$
Avogadro's Number	N_A	$6.02214199(47) \times 10^{23} \, mol^{-1}$
Bohr Magneton	μ_B	$9.27400899(37) \times 10^{-24} \, J \, T^{-1}$
Bohr Radius	a_0	$0.5291772083(19) \times 10^{-10} \, m$
Boltzmann's Constant	k	$1.3806503(24) \times 10^{-23} \, J \, K^{-1}$
Compton Wavelength	λ_c	$2.426310215(18) \times 10^{-12} \, m$
Deuteron Mass	m_d	$3.34358309(26) \times 10^{-27} \, kg$
Electric Constant	ε_0	$8.854187817 \times 10^{-12} \, F \, m^{-1}$
Electron Mass	m_e	$9.10938188(72) \times 10^{-31} \, kg$
Electron-Volt	eV	$1.602176462(63) \times 10^{-19} \, J$
Elementary Charge	e	$1.602176462(63) \times 10^{-19} \, C$

Fig. 16.1 The universe as we know it would not exist—and we would not be here—if certain (most probably all) of its physical constants were not fine-tuned (mostly within 1 to 2 percent) to their current values. In the picture, some of these constants are displayed. A more complete list—and exemplary descriptions of what would happen to the universe if some were slightly different—can be found in Biocentrism by R. Lanza and B. Berman (BenBella, 2010).

presence and how he or she makes a measurement unambiguously determine what a physical object becomes. Measure it at one spot and an electron is a wave. Enter the picture a bit sooner, bringing our awareness onto the scene at an intermediate point—the slit rather than the final detection point, say—and the electron lives its life as a particle. An open-and-shut case.

A SPECIAL CASE: CHANGING c, $\hbar$, G, AND ε_0

The constants c (the speed of light), $\hbar$ (the reduced Planck constant), G (the gravitational constant), and ε_0 (the dielectric constant) are *the fundamental constants* in the sense that their values can be chosen arbitrarily. In other words, there exist systems of units in which those four constants have arbitrary values, while other physical quantities and measured constants are then given as multiples of the units, defined in terms of c, $\hbar$, G, and ε_0. An example of such a system is the famous *Planck system of units* in which $c = \hbar = G = 1$, and its extension, in which $c = \hbar = G = 4\pi\varepsilon_0 = 1$.

The unit of length, namely the meter, is currently defined in terms of the speed of light—which is ascribed or assigned a fixed value close to 3×10^8 m/s. Thus, nowadays the numerical value of the speed of light is *defined as being fixed*. That is, it is no longer considered a measurable quantity. (For more, see the Wikipedia article for "Meter.")

The speed of light enters the equation for the constant alpha, which determines the strength of the electromagnetic interaction: $\alpha = e^2/(4\pi\varepsilon_0 \hbar c)$.

As you can see, changing c, while leaving the other three fundamental constants fixed, would change α as well, and consequently all of atomic physics, including the possibility of life as we know it.

On the other hand, one can change c and at the same time also change ε_0, $\hbar$, and G, so that a remains the same. In such a way we would not have changed physics, but only the units in which physical quantities are expressed.

Of course, among the two hundred or so constants/parameters, there might be some whose values are not of decisive importance for the making of our universe. However, it is very likely that behind those two hundred constants/parameters, there is an underlying fundamental theory, a relationship that explains them all. If so, changing any one of them would change the very structure of the universe.

As we look for ways to offer science the hard evidence it craves, another possible strategy is to investigate when and how time—evolution time—starts to exist. Yes, this is a mind-twisting idea, but if "time" is our label for a before-and-after sequence of events, then the physical unfolding of measurable consequences cannot happen absent time. And if, as discussed earlier in this book, observers' minds with their ability to remember the past supply the vital mechanism needed for memory and thus comparisons, time provides us with a perfect illustration of biocentrism's necessity.

We've already heard the 2,500-year-old tale of Zeno of Elea, who reasoned that an arrow must be in only one location during any given instant of its flight. But, he continued, if the arrow is in only one place, it must be, however momentarily, at rest. At every moment of its trajectory, the arrow must be present somewhere, at some specific location. Logically, then, what is occurring is not motion per se: rather, it is a series of separate events. Likewise, the forward motion of time—of which the movement of the arrow is an embodiment—is not a feature of the external world but a projection that arises within us, as we tie together events we are observing.

Time has no meaning without relationship to another point. It is a relational concept, one event relative to another. Thus, to have an arrow or directionality of time, there *must* be an observer with memory. We're back to the inescapability of the conscious observer.

Speaking of time, however, it's time now to wrap up the recap of our revelations and, just as importantly, explore how they might change our perception of our own lives, our future, and the very nature of our daily reality. No reader can be blamed for desiring further clarification of how quantum theory (popularly renowned for being esoterica personified), or the details of how observers cause subatomic particles to behave, might apply concretely to their own lives.

Those who are somewhat—but perhaps not completely—clear on what biocentrism is, what it reveals, and the evidence behind it have several options as we near the end of this book. First, pay attention to the overview in these final few pages. Second, if you're scientifically or mathematically inclined, you can turn to the appendixes, where some of the "hard science" evidence is reproduced in full. You might, too, enjoy our presentation (also in an appendix) of some of the common questions raised by biocentrism's critics, and our answers and rebuttals—which may well answer any remaining queries of your own.

But first, let's look again at all eleven of biocentrism's principles. If any particularly intrigue you, be aware that the first seven each had their own chapter in the first *Biocentrism* book, with leisurely explanations accompanied by illustrations; the final four are derived from material in the book before you now.

After restating the principles, we'll look at what they mean in terms of the cosmos, life in general, and the lives we live as individuals.

PRINCIPLES OF BIOCENTRISM

First principle of biocentrism: What we perceive as reality is a process that involves our consciousness. An external reality, if it existed, would by definition have to exist in the framework of space and time. But space and time are not independent realities but rather tools of the human and animal mind.

Second principle of biocentrism: Our external and internal perceptions are inextricably intertwined. They are different sides of the same coin and cannot be divorced from one another.

Third principle of biocentrism: The behavior of subatomic particles —indeed all particles and objects—is inextricably linked to the presence of an observer. Absent a conscious observer, they at best exist in an undetermined state of probability waves.

Fourth principle of biocentrism: Without consciousness, "matter" dwells in an undetermined state of probability. Any universe that could have preceded consciousness only existed in a probability state.

Fifth principle of biocentrism: The structure of the universe is explainable only through biocentrism because the universe is fine-tuned for life—which makes perfect sense as life creates the universe, not the other way around. The "universe" is simply the complete spatiotemporal logic of the self.

Sixth principle of biocentrism: Time does not have a real existence outside of animal sense perception. It is the process by which we perceive changes in the universe.

Seventh principle of biocentrism: Space, like time, is not an object or a thing. Space is another form of our animal understanding and does not have an independent reality. We carry space and time around with us like turtles with shells. Thus, there is no absolute self-existing matrix in which physical events occur independent of life.

Eighth principle of biocentrism: Biocentrism offers the only explanation of how the mind is unified with matter and the world by showing how modulation of ion dynamics in the brain at the quantum level allows all parts of the information system that we associate with consciousness to be simultaneously interconnected.

Ninth principle of biocentrism: There are several basic relationships—called "forces"—that the mind uses to construct reality. They have their roots in the logic of how the various components of the information system interact with each other to create the 3-D experience we call consciousness or reality. Each force describes how bits of energy interact at different levels, starting with the strong and weak forces (which govern how particles hold together or fall apart in the nucleus of atoms) and moving up to electromagnetism and then gravity (which dominates interactions on astronomical scales such as the behavior of solar systems and galaxies).

Tenth principle of biocentrism: The two pillars of physics—quantum mechanics and general relativity—can only be reconciled by taking observers, us, into account.

Eleventh principle of biocentrism: Observers ultimately define the structure of physical reality—of states of matter and spacetime—even if there is a "real world out there" beyond us, whether one of fields, quantum foam, or some other entity.

* * *

After looking these principles over for a final time, the reader might understand and even be excited by them while still not fully following these threads to where they are woven into our own lives. So let's take a deeper dive into the implications.

Let's assume the newly published science discussed in this book and reproduced in its appendixes is merely the start of serious investigations that ultimately make biocentrism the global standard model of how the universe works. Say it becomes the accepted scientific reality, the worldview that accounts for how most people regard the cosmos and their place in it. What would that really mean?

First and foremost, it would mean that the fundamental ground state of the universe is not empty space, nor dumb, randomly colliding particles. Instead, that view would be replaced with the knowledge that the basis of the universe is conscious life. Which itself, though we haven't bothered to spell this out in so many words, is infused with exquisite underlying intelligence. In other words, this would mean that the cosmos is not senseless, and if this isn't good news, what is?

It would also mean that the supposed yawning endless emptiness of the cosmos is not real. I'm guessing you will happily accept this development, too. Who among us is attached to nothingness?

So: the Lonely Hearts Club aspect of the cosmos vanishes. And the big bang, that classical-science "explanation" for the genesis of everything, reverts to a hollow, meaningless oddity, a non-clarification—maybe not such a surprise, since the notion of everything arising mysteriously from "nothing" never seemed like a thesis any teacher would award with a passing grade.

Loren Eiseley, the great naturalist, once said that scientists "have not always been able to see that an old theory, given a hairsbreadth twist, might open an entirely new vista to the human reason." Cosmic evolution turns out to be the perfect case of this. Amazingly, it all makes sense if you assume that the big bang is the end of the chain of physical causality, not the beginning. The observer is the first cause, the vital force that collapses not only the present, but the cascade of spatiotemporal events we call the past. Stephen Hawking was right when he said: "The past, like the future, is indefinite and exists only as a spectrum of possibilities."

Next, "mind" or "consciousness" becomes the essence or matrix of the cosmos, which, again, means that life is central to everything. Talk about "beginnings" loses all urgency, since time never existed outside of consciousness to begin with.

Speaking of which, if consciousness is everywhere and never discontinuous, then there's no death to experience. Sure, that dead dog in the road isn't going to get back up and again put his muddy paws on your pants. But in terms of awareness, you have never not experienced consciousness and its myriad sense impressions, nor will this parade ever cease. You can count on this. So, biocentrism has handed you the "no death" card—it's unlikely you'll ever want to trade it in again for something else. If you're bummed out by the fact that your experiences may not always be witnessed through your present eyes in your present body, well, you get what you pay for.

As a further bonus, once you've truly understood that all experiences occur strictly in the mind, so that the blue skies and pretty flowers you see are not physically apart from you "out there," the ensuing sense of oneness often produces a profound peace and serenity. Whether "peace of mind" is something you've personally coveted or not, many attest that it is a worthy goal.

Finally, of course, there is the alluring dance of future possibilities. With time and space firmly recognized as being "internal" properties of your own perceptions, biocentric technological developments may well allow travel through time, in ways that would be impossible if those dimensions were true external barriers.

But above and beyond all this, acceptance of biocentrism would give us not only a worldview that unites us all more intimately than could be achieved by any government program, but a scientific model that—incorporating the centuries of hard-won breakthroughs outlined in this book—at last makes sense.

POST SCRIPTUM:
THE MAN WHO CARED

Sometimes, a problem—whether a personal matter or one of science—seems insoluble due to inertia or a simple unwillingness to flexibly evaluate a new circumstance. That's where physics found itself before the start of the First World War—a logjam finally broken by a small cadre of rule breakers. To one of the authors, Robert Lanza, this example mirrors his own predicament a half century ago, which was also resolved by a hero.

James Watson—who discovered the DNA double helix—once remarked, "You've got to be prepared sometimes to do some things that people say you're not qualified to do." He also said, "Since you know you're going to get into trouble, you ought to have someone to save you after you're in deep shit. So you better always have someone who believes in you."

For me, that someone was Eliot Stellar, the provost of the University of Pennsylvania, and chair of the Human Rights Committee of the prestigious National Academy of Sciences.

I was in trouble at various times as a student, but that never stopped me from forging ahead along whatever path had led me into peril, because I knew Eliot Stellar would save me. I was young and idealistic—I was not only unhappy with how science described the world, but also unhappy with its failure to use the available achievements and know-how to improve the human condition in large parts of the world.* When I was in medical school, I decided to compile a book that I hoped would address the latter of these concerns by offering a multifaceted picture of where medicine and science stood and where it intended to go, made up of contributions from leading scientists from various disciplines discussing the state of the science and their thoughts and suggestions for necessary changes in the future.

To choose among the many possible contributors was not easy, and I was not at all sure what their reaction would be to my request. In the end, I wrote to heart transplant pioneer Christiaan Barnard, along with the US surgeon general, the director-general of the World Health Organization, and recipients of the Nobel and Lenin Peace prizes, among others. The response was overwhelming and gratifying, dispelling any doubt I may have had about the need for and relevance of the kind of assessment and commentary that I intended the book to offer. But it was this response that gave rise to the problem.

You see, I had used my medical school mailbox address on the invitation letters. The dean's office began to receive telephone calls trying to locate me . . . from the US surgeon general, for instance. This outraged the dean of students, who wanted me to send out follow-up letters, explaining to the individuals who received a request from me to contribute to the book that I was a medical student. In his mind, there was a risk that the project could fail, and thus upset a lot of very important people. He was right, of course.

However, I believed that to send out such letters would undermine the confidence of my hoped-for contributors; more importantly, to my mind, the book was my own personal project, and thus none of the dean's business. And when he called me into his office and ordered me to send out the follow-up letters, I said just that. In response to my refusal, he threatened me, telling me that if I didn't do as I was told, I wouldn't

---

* In a job reference for me, Stellar once said, "He is a bit of a renegade, but so was Einstein." I'm not sure the comparison to Einstein was deserved, but my reputation as a renegade or troublemaker certainly was.

receive my MD degree. At which point, I told him that I'd already gotten what I'd come there for—a medical education. I hadn't come there for a piece of paper, I told him. He seemed taken aback.

The conversation became heated, and finally, the dean said, "I've never had a student talk to me like this!" I stood up and pointed my finger straight into his eyes and said, "I'm talking to you as one human being to another." We were making quite a bit of noise, and just then, there was a knock on the door and a voice said, "Fred, is everything okay? We're late for our meeting."

"I'm going to be late, go on without me," the dean replied. Ending our confrontation, he told me that I had better have a faculty advisor to defend me.

Of course, I went straight to Eliot Stellar and explained things. "Who is your advisor?" he asked me. I replied that I didn't have one. He leaned back in his chair and seemed a bit puzzled. Finally, he said, "I suppose it's okay to be your own advisor."

The next day I was summoned back to the dean's office. This time, the dean greeted me with a warm smile and said, "You should have told me that Eliot Stellar was your advisor."

I still refused to comply with the dean's demands, however, and he called me before the Student Standards Committee. Things with the committee went pretty much as they had with the dean—that is, badly. They sent me a letter, stating:

> Be advised that if you do not fulfill the Student Standards Committee required course of action, the recommendation open to the subcommittee is to decline recommending you for graduation. The potential sanctions which might be imposed include, but are not limited to, suspension or dismissal. Because of the serious nature of the questions being raised by the Student Standards Committee and because you are in jeopardy of being dismissed from the School of Medicine . . . I would recommend that you meet with your faculty advisor, Dr. Eliot Stellar, to be certain you understand the ramifications of your position.

I was in deep s---.

But Eliot Stellar stood behind me. "You shouldn't be in this all alone," he said.

I clung to my position in the months that followed, and the severity of my defiance continued to displease the dean and the Student Standards Committee.

"They're bureaucrats," Dr. Stellar explained. "They just don't understand." The sixties had ended a decade ago, but he still valued and fought for the individuality and creativity that period had embraced.

I have always been convinced that if it hadn't been for Dr. Stellar working behind the scenes, I never would have graduated from medical school. I never would have become a doctor. One night, after I had sent a particularly provocative letter to the dean, Eliot Stellar called me at home. He was trying to put out the fires my stubbornness had started, and he asked that I hold off on sending any more letters to the dean without checking with him. I had worked hard, he told me during the conversation; I had earned an MD degree.

"The degree isn't important," I said. "I got what I came here for—a medical education."

And at about that point I heard his wife, Betty, say in the background, "Tell him to ask his mother!"

"Shhh!" said Eliot. "It's his decision."

I seemed to have no allies except Eliot, and I frequently went to him when the going got rough. The day a settlement was reached, I was in his office. While we were talking, the phone rang. After listening to the person in silence for a minute or two, he finally said to the caller, "The emergency is off." Afterward, I thanked him for caring, for not joining forces with the office of the dean.

"I would like to think," he said, "that I made things a little more fair."

* * *

A few years later, I boarded a trolley car into the city and took an empty seat next to a well-dressed woman. After a few minutes, she turned to me: "You're Robert Lanza, aren't you?" Yes, I said—why? The woman replied that she had worked in the dean's office, and she remembered the day of my fight with the dean well. All the office staff had been standing outside the door listening, she told me—and they all cheered silently when I told him off.

The book I compiled, *Medical Science and the Advancement of World Health*, was published in 1985. The dedication reads: "To Eliot Stellar—For the inspiration of his human kindliness and his virtuous and enlightened life, as well as for the courage and insight in creating the University Scholars Program at the University of Pennsylvania, which introduces changes in the educational system that nurture creativity and personal growth—changes that are essential if future generations are to cope successfully with the challenges that threaten humankind's very existence."

If my tone in the telling of this story seems detached, it is because this is a tribute to Eliot Stellar, who once told me, "Let the facts speak for themselves." Eliot Stellar, my mentor, one of the greatest physiological psychologists ever to live, and arguably the most decent human being I ever met, died in 1993.

I miss him.

Many years after I graduated, I ran into the dean in the hallway. He shook my hand and said, "As one human being to another" (referring, of course, to the day I said the same to him in his office). He then congratulated me on all I had accomplished since I graduated. Eliot Stellar would have been very happy to see it.

APPENDIX 1:
QUESTIONS AND
CRITICISMS

Question: If consciousness created reality, then where did consciousness come from?

In response to a 2007 Q&A with one of the authors (Lanza) published in *Wired*,* science writer Adam Rogers wrote a follow-up blog post, in which he stated, "Lanza's conclusion is that we need to understand the mysteries of consciousness so we can explain how individual clumps of neurons produce—from what he does not say—little slivers of illusory universe. Bit of a chicken-and-egg problem there, I think. Those neurons might not be the end of the story on how consciousness is produced (another question from What We Don't Know, I might add) but they are at least the beginning."

Response:

The supposed "chicken-and-egg problem" doesn't exist. Rogers is looking at the new paradigm through the eyes of the old one. Time is not "out

---

* Aaron Rowe, "Will Biology Solve the Universe?" *Wired*, March 8, 2007.

there" ticking away like a clock. "Before" and "after" have no absolute meaning independent of the observer. Thus the question of what came before consciousness is meaningless, and only arises due to an incomplete understanding of physics. The world we perceive is defined by us (see Chapters 11 and 14).

Question: Is there a difference between the physical brain and the mind?

One widely cited critique of biocentrism, published at nirmukta.com, states, "How can the 'living, biological creature' exist if the universe has not been created yet? It becomes apparent that Lanza is muddling the meaning of the word 'consciousness.' In one sense he equates it to subjective experience that is tied to a physical brain. In another, he assigns to consciousness a spatiotemporal logic that exists outside of physical manifestation."

Response:

Biocentrism shows that the external world is actually within the mind— not "within" the brain. The brain is an actual physical object that occupies a specific location. It exists as a spatiotemporal construction. Other objects, like tables and chairs, are also constructions and are located outside the brain. However, brains, tables, and chairs alike all exist in the "mind." The mind is what generates the spatiotemporal construction in the first place. Thus, the mind refers to pre-spatiotemporal, and the brain to post-spatiotemporal. You experience your mind's image of your body, including your brain, just as you experience trees and galaxies. The mind is everywhere. It is everything you see, hear, and sense. The brain is where the brain is, and the tree is where the tree is. But the mind has no location. It is everywhere you observe, smell, or hear anything.

Question: In what sense is biocentrism a theory? Can biocentrism be falsified?

Several critics claim biocentrism, like string theory, is unfalsifiable (that is, it cannot be disproved) and therefore cannot properly be considered a scientific theory.

Response:

This is patently false. Biocentrism can be tested using a range of different experiments—for instance, scaled-up superposition. Indeed, the observer-linked variations described in Podolskiy, Barvinsky, and Lanza's newest work (see Chapter 14 and Appendix 3) are testable. They can be checked by performing both real and numerical experiments on various quantum-mechanical systems. In fact, the results have already been checked numerically and will be checked experimentally in the near future.

Indeed, yet another biocentric prediction was experimentally confirmed as this book was being written. Massimiliano Proietti and his colleagues at Heriot-Watt University in Edinburgh performed a quantum experiment showing there is no such thing as objective reality ("Experimental test of local observer independence," *Science Advances,* September 20, 2019). Physicists had long suspected that quantum mechanics allows two observers to experience different, conflicting realities. "If one holds fast to the assumptions of locality and free choice," wrote the authors, "this result implies that quantum theory should be interpreted in an observer-dependent way."

Future experiments along these lines are likely to test additional tenets of biocentrism. But biocentrism's adherents are unlikely to be surprised by the results. As Eugene Wigner himself once said, "The very study of the external world [leads] to the conclusion that the content of consciousness is an ultimate reality."

Question: Biocentrism claims that the colors we see only exist in our head. But how can that be true if light particles exist in the external world that correspond to these various colors?

Nirmukta states it this way:

> *If you dig into what Lanza says it becomes clear that he is positioning the relativistic nature of reality to make it seem incongruous with its objective existence. His reasoning relies on a subtle muddling of the concepts of subjectivity and objectivity. Take, for example, his argument here:*
> *"Consider the color and brightness of everything you see 'out there.' On its own, light doesn't have any color or brightness at all. The*

unquestionable reality is that nothing remotely resembling what you see could be present without your consciousness. Consider the weather: We step outside and see a blue sky—but the cells in our brain could easily be changed so we 'see' red or green instead. We think it feels hot and humid, but to a tropical frog it would feel cold and dry. In any case, you get the point. This logic applies to virtually everything."

There is only some partial truth to Lanza's claims. Color is an experiential truth—that is, it is a descriptive phenomenon that lies outside of objective reality. No physicist will deny this. However, the physical properties of light that are responsible for color are characteristics of the natural universe. Therefore, the sensory experience of color is subjective, but the properties of light responsible for that sensory experience are objectively true. The mind does not create the natural phenomenon itself; it creates a subjective experience or a representation of the phenomenon.

Response:

Nirmukta's argument is flawed on multiple levels. The "properties" of any photon or bit of electromagnetic radiation are wavelength and frequency —meaning, the oscillations of magnetic and electric fields. Visible light accounts for only a small portion of the electromagnetic spectrum, which is a continuous gradient that runs from shorter to longer wavelengths and includes gamma rays, radar, radio, and microwaves (none of which we perceive as "color"). Such fields are not "responsible" for the perception of color; indeed, they themselves are wholly invisible. At best, we should experience the visual spectrum as nothing but a grayscale continuum ranging from dark to light; it should by all rights be a simple quantitative experience. However, that is not the case. Instead we have a unique qualitative experience that we subjectively experience as distinct colors when light falls within very specific ranges of the visual spectrum (see Chapter 7).

In truth, the "responsibility" for, or cause of, colors lies with the way the animal mind reacts to invisible energies by creating the experience of, say, "red" or "blue." And, indeed, on a more fundamental level, these photons themselves only arise upon observation and wave function collapse; experiments clearly show that particles of light themselves do not exist with real properties until they are actually observed.

None of this is controversial. The fact that colors do not exist "out there" on their own has been well established for centuries, as evidenced by Isaac Newton's assertion in *Opticks* that "the rays . . . are not coloured." As Canadian physicist Roy Bishop writes in each annual edition of his popular *Observer's Handbook*, "The eye does not detect the colors of the rainbow; the brain creates them."

Question: What about all the evidence documenting the evolution of life and the universe?

Nirmukta asks, "Can Lanza deny all the evidence that, whereas we humans emerged on the scene very recently, our Earth and the solar system and the universe at large have been there all along? What about all the objective evidence that life forms have emerged and evolved to greater and greater complexity, resulting in the emergence of humans at a certain stage in the evolutionary history of the earth? What about all the fossil evidence for how biological and other forms of complexity have been evolving? How can humans arrogate to themselves the power to create objective reality?"

Response:

The question is how to interpret this "evidence" in terms of physical reality—that is, whether we should continue to cling to the old deterministic framework.

Although classical evolution does an excellent job of helping us understand the past, it fails to capture evolution's driving force. To do so, evolution needs to add the observer to the equation.

Many believe that the universe was, until fairly recently, a lifeless collection of particles bouncing off one another, existing and unfolding without us. It's presented as a watch that somehow wound itself up, and that will unwind in a semi-predictable way. But it is we observers who create the arrow of time (see Chapter 11). As Stephen Hawking stated, "There is no way to remove the observer—us—from our perceptions of the world . . . In classical physics, the past is assumed to exist as a definite series of events, but according to quantum physics, the past, like the future, is indefinite and exists only as a spectrum of possibilities."

If we, the observer, collapse these possibilities (that is, the past and the future), then where does that leave evolutionary theory, as described in our schoolbooks? Until the present is determined, how can there be a past? The fact is, the universe does not run mechanistically like a clock, independent of us, and it never has. The past begins with the observer, not the other way around.

Nirmukta asks, "What about all the fossil evidence?" but fossils are really no different than anything else in nature. The carbon atoms in your body, for instance, are "fossils" created in the heart of exploding supernova stars. The bottom line is that *all* physical reality begins and ends with the observer. "We are participators," Wheeler said, "in bringing about something of the universe in the distant past." The observer is the first cause, the vital force that collapses not only the present but the cascade of past spatiotemporal events we call evolution.

Question: Can we change the world around us with "mind powers"?

In response to an article one of the authors (Lanza) published in the *Humanist*,[†] physicist Victor Stenger wrote: "The world would be a far different place for all of us if it were just all in our heads—if we really could make our own reality as the New Agers believe. The fact that the world rarely is what we want it to be is the best evidence that we have little to say about it. The myth of quantum consciousness should take its place along with gods, unicorns, and dragons as yet another product of the fantasies of people unwilling to accept what science, reason, and their own eyes tell them about the world."

Response:

Biocentrism does not in any way hold that we can simply "make our own reality" according to our specifications. In the *Wired* interview referred to earlier in this appendix, the interviewer asked, "Do you expect that some people will read your article and think you mean that they can sit on a mountaintop and meditate to change the world around them with mind powers?" Lanza replied, "We can't decide that we want to jump off the

---

† "The Wise Silence," November/December 1992.

roof and not get hurt. However much we want, we can't violate the rules of spatiotemporal logic."

If you go to the grocery store and buy a box of cornflakes or Grape Nuts, you will not find Froot Loops in your cupboard the next morning—no matter how much you may want them.

Question: Copenhagen or many-worlds interpretation?

One reviewer, referring to *Biocentrism* (page 58), wrote:

> You say: "If we want some sort of alternative to the idea of an object's wave-function collapsing just because someone looked at it, and avoid that kind of spooky action at a distance, we might jump aboard Copenhagen's competitor, the 'Many Worlds Interpretation' (MWI), which says that everything that can happen, does happen . . . According to this view, embraced by such modern theorists as Stephen Hawking, our universe has no superpositions or contradictions at all." But you add, "All the entangled experiments of the past decades point increasingly toward confirming Copenhagen more than anything else. And this, as we've said, strongly supports biocentrism." Which view do you find most persuasive? And how would it change biocentrism if one view or the other were correct?

Response:

According to biocentrism, the Copenhagen interpretation is more or less correct, but requires important modifications:

- Physical systems do not have definite properties prior to being measured—*and* wave function collapse only occurs with measurements by a living observer, not measurements made by an inanimate object such as a camera or other measuring device that records information (see next question). The latter information exists in superposition until observed by a consciousness.
- The wave function that collapses is not a "real" thing—it is merely a statistical interpretation.

- Superposition is not a "real" thing—it represents statistical possibility.

The general idea of "many worlds" and a "multiverse" is also compatible with biocentrism. Unfortunately, several key parts of the formal MWI are likewise in need of modification:

- Most versions of the many-worlds interpretation include this idea: that the equations of physics that model the time evolution of systems without embedded observers are sufficient for modeling systems that do contain observers; in particular, there is no purely observation-triggered wave function collapse of the sort the Copenhagen interpretation proposes. Of course, according to biocentrism, this is incorrect.
- All "possible" alternative histories and futures are indeed real, each representing an actual world/universe. However, it is extremely important to point out that no world or universe can exist independent of a conscious observer.
- The starting "universal wave function" does not have an objective reality—it is merely a statistical description of the possibilities.

Question: Does decoherence/wave function collapse require a conscious observer?

"Evolution does not need an observer," claims Steven Novella, an assistant professor of neurology at Yale best known for his skeptical blog posts. "There is nothing in the process of evolution, and no observation of nature that requires it. Bohr is talking about a quantum phenomenon of the collapse of the probability wave. But this does not require a literal observer, just interaction with the surrounding environment . . . the universe can observe itself just fine without us."

Response:

Some scientists believe that just encountering another particle will collapse a particle's wave function—that is, the environment itself can do this.

But others, like us, think it takes something much more macroscopic—in fact, a living observer—to decohere a probabilistic quantum state.

We know that not every measurement leads to the loss of quantum coherence (that is, wave function collapse). Elementary particles in the subatomic realm, for instance, do not lose quantum coherence despite the fact that they probe each other's states all the time. In order for wave function collapse to occur, the device measuring the state of a quantum object has to be macroscopic. This, it seemed for a long time, explained why the physics of the microscopic world is so drastically different from the physics of the events and objects that surround us in everyday life.

Why does collapse occur when the device or object doing the observing is macroscopic? Being "macroscopic" means that not all parts of the object are being observed at once, and so their properties are unknown. Such incompleteness is well known to cause decoherence and wave function collapse.

For example, if we have two electrons in an entangled state, measuring the properties of only one electron without having information about the second particle will lead to an apparent decoherence, a breakdown of the entanglement of the two particles. On the other hand, if you gain information about the states of both entangled particles, experiments show that the entanglement of the two particles is reestablished.

If you could measure the quantum states of all the particles in the universe simultaneously, you would never experience the deterministic world we live in, where everyone is either alive or dead, but only the probabilistic blur of quantum mechanics. But, of course, the world is deterministic for us, simply because of how our senses and brains work. For instance, our eyes cannot detect ultra-high-energy cosmic rays, the cosmic microwave background radiation, or the tiny motions of sub-atomic particles. Our senses are limited, and our brains cannot process all events simultaneously happening in the universe. Ultimately, since we are unable to see and perceive the complete universe, we experience its state as one with quantum coherence seemingly lost.

Our paper "On Decoherence in Quantum Gravity" (see Appendix 2) clearly shows that the intrinsic properties of quantum gravity and matter alone cannot explain the tremendous effectiveness of the emergence of time and the lack of quantum entanglement in our everyday world. Quantum gravitational decoherence is too ineffective to guarantee the emergence of the arrow of time and the "quantum-to-classical" transition that happens at scales of physical interest. Our paper argues that the emergence of the arrow of time is directly related to the nature and properties of the physical observer; a "brainless" observer does not experience time and/or decoherence with any degree of freedom.

A Final Discussion

In Spring 2007, one of the authors (Lanza) introduced biocentrism in an essay in the *American Scholar* titled "A New Theory of the Universe—Biocentrism Builds on Quantum Physics by Putting Life into the Equation." Astrophysicist and science writer David Lindley published a response in *USA Today*:[‡]

> *I take issue with [Lanza's] views about physics. He wants to argue that all of physical reality is in our mind but his interpretations of relativity and quantum mechanics are misguided.*
>
> *First, he claims that Einstein made space and time observer-dependent and thus subjective, so that there's no such thing as space and time except insofar as we perceive them. I don't agree. It's true that Einstein got rid of the old Newtonian absolutes, and showed that measurements of space and time are not the same for all observers. But—and this is crucially important—he constructed a new system of spacetime that shows how such differing measurements can be reconciled. That is, relativity retains an objective physical framework called spacetime, with a specific geometrical structure, but it allows observers to map out spacetime in different ways.*

Lindley adds: "Lanza takes off with the idea that you need consciousness to 'create' reality. This view has had some supporters over the years, but it's always been an odd attitude, and today is not taken seriously."

---

‡ "Exclusive: Response to Robert Lanza's Essay," *USA Today*, March 8, 2007.

He ends by saying: "Finally, I can't help thinking that there's an enormous exercise of vanity in Lanza's argument—the universe only exists, he says, because we're here to observe it and be part of it. I would go the opposite extreme. I think the universe was a real physical thing long before we came on the scene, and we humans are just crumbs of organic matter clinging to the surface of one tiny rock. Cosmically, we are no more significant than mold on a shower curtain."

Lanza's Response: David Lindley misrepresents and oversimplifies the biocentric position throughout his article. For example, he states that I claim that Einstein made space and time subjective. This is simply false. The spacetime Einstein conceived of in his theory of special relativity is an independent reality having its own existence and its own structure. It is a "clockwork" that ticks away regardless of whether an observer is present. It has just as much reality for an inanimate object like a planet or a star as for a living creature like a woodchuck or a human being. Einstein's theory ascribes objective reality to spacetime independent of the occupation of whatever events happen to take place in its arena. It is only with hindsight that we realize that Einstein merely substituted a 4-D absolute entity for a 3-D one. In fact, at the beginning of his paper on general relativity, Einstein raised the same concern about his theory of special relativity.

Physicists believe that they can build from nature without including the living. But if indeed there is a place for science to set its foundation safely, it is not where they imagine. Physicists, of course, are obsessed with mathematics and equations, black holes and particles of light. As a consequence, they miss much of what's just outside their window. They live in a cloud above the world. However, the ducks and the cormorants—paddling out there in the pond beyond the lily pads and cattails—the butterfly and the wolf; they are all an important part of the answer. Many in science have yet to learn that the universe cannot be disparted from the life that lives within its walls.

Lindley also quotes a line from my essay that reads "the kitchen disappears when we're in the bathroom," responding: "How can this be? Are we really supposed to imagine that the kitchen goes away when we're not in it, and returns in the exact same form when we come back in?" Of course, the wave function of the kitchen collapses when we first observe it, and a record of this collapse remains in our memory.

Finally, Lindley says, "Lanza takes off with the idea that you need consciousness to 'create' reality . . . it's always been an odd attitude and today is not taken seriously." It may not be taken seriously by Lindley, but it certainly is by many. Werner Heisenberg—Nobel laureate and founder of quantum mechanics itself—said, "Contemporary science, today more than at any previous time, has been forced by nature herself to pose again the old question of the possibility of comprehending reality by mental processes, and to answer it in a slightly different way." Indeed, Eugene Wigner, another of the twentieth century's greatest physicists, stated that it is "not possible to formulate the laws of [physics] in a fully consistent way without reference to the consciousness [of the observer]."

Should you want more recent examples, consider again the provocative 2007 experiment that was published in *Science* (see Chapter 7).§ This landmark experiment showed that a choice made now can influence an event retroactively—an event that has already occurred in the past. This and other experiments clearly show that space and time are relative to the observer. Experiments also continue to show that the properties of matter itself are observer determined. In these experiments, a particle goes through one hole if you look at it, but if you don't look at it, it instead passes through more than one hole at the same time. Science has so far offered no explanation for how the world can be like that. The theory I'm proposing, of life and consciousness-centered reality, is the first that offers a scientifically cogent account.

We need to confront the experiments that have accumulated. We can't continue to say only "gee, that's weird" and then put our heads firmly back in the sand. The goal of science is to explain the world around us. Yet, despite all the evidence, scientists continue to regard the observer as an inconvenience, observer-linked effects as an oddity gumming up their theories. Our theory ascribes the *answer* to the observer—to the

---

§ Jacques et al., "Experimental Realization of Wheeler's Delayed-choice *Gedanken* Experiment." *Science* 315, 966 (2007).

biological creature—rather than to matter. And through it, for the first time, all of the bizarre findings of relativity and quantum theory make sense.

The physicists at the helm have failed to reconcile the foundations of science for over a hundred years. It's time to open up the discussion about the nature of the universe, not only to the entire scientific community, but to all of society.

It's time for a rethink.

APPENDIX 2:
THE OBSERVER AND
THE ARROW OF TIME

Non-technical summary of paper:
In his papers on relativity (which were also published in this journal), Einstein showed that time was relative to the observer. This new paper takes this one step further, arguing that the observer creates it. The paper shows that the intrinsic properties of quantum gravity and matter alone cannot explain the tremendous effectiveness of the emergence of time and the lack of quantum entanglement in our everyday world. Instead, it's necessary to include the properties of the observer, and in particular, the way we process and remember information.

On decoherence in quantum gravity

Dmitriy Podolskiy[1],* and Robert Lanza[2]

Received 13 January 2016, revised 24 June 2016, accepted 27 July 2016
Published in *Annalen der Physik* 528, 663-676, 2016

It was previously argued that the phenomenon of quantum gravitational decoherence described by the Wheeler-DeWitt equation is responsible for the emergence of the arrow of time. Here we show that the characteristic spatio-temporal scales of quantum gravitational decoherence are typically logarithmically larger than a characteristic curvature radius $R^{-1/2}$ of the background space-time. This largeness is a direct consequence of the fact that gravity is a non-renormalizable theory, and the corresponding effective field theory is nearly decoupled from matter degrees of freedom in the physical limit $M_P \to \infty$. Therefore, as such, quantum gravitational decoherence is too ineffective to guarantee the emergence of the arrow of time and the "quantum-to-classical" transition to happen at scales of physical interest. We argue that the emergence of the arrow of time is directly related to the nature and properties of physical observer.

1 Introduction

Quantum mechanical decoherence is one of the cornerstones of the quantum theory [1, 2]. Macroscopic physical systems are known to decohere during vanishingly tiny fractions of a second, which, as generally accepted, effectively leads to emergence of a deterministic quasiclassical world which we experience. The theory of decoherence has passed extensive experimental tests, and dynamics of the decoherence process itself was many times observed in the laboratory [3–15]. The analysis of decoherence in non-relativistic quantum mechanical systems is apparently based on the *notion of time*, the latter itself believed to emerge due to decoherence between different WKB branches of the solutions of the Wheeler-DeWitt equation describing quantum gravity [2, 16–19]. Thus, to claim understanding of decoherence "at large", one has to first understand decoherence in quantum gravity. The latter is clearly problematic, as no consistent and complete theory of quantum gravity has emerged yet.

Although it is generally believed that when describing dynamics of decoherence in relativistic field theories and gravity one does not face any fundamental difficulties and gravity decoheres quickly due to interaction with matter [20–23], we shall demonstrate here by simple estimates that decoherence of quantum gravitational degrees of freedom might in some relevant cases (in particular, in a physical situation realized in the very early Universe) actually be rather ineffective. The nature of this ineffectiveness is to a large degree related to the non-renormalizability of gravity. To understand how the latter influences the dynamics of decoherence, one can consider theories with a Landau pole such as the $\lambda\phi^4$ scalar field theory in $d = 4$ dimensions. This theory is believed to be trivial [24], since the physical coupling λ_{phys} vanishes in the continuum limit.[1] When $d \geq 5$, where the triviality is certain [25, 26], critical exponents of $\lambda\phi^4$ theory and other theories from the same universality class coincide with the ones predicted by the mean field theory. Thus, such theories are effectively free in the continuum limit, i.e., $\lambda_{\mathrm{phys}} \sim \frac{\lambda}{\Lambda^{d-4}} \to 0$ when the UV cutoff $\Lambda \to \infty$. Quantum mechanical decoherence of the field states in such QFTs can only proceed through the interaction with other degrees of freedom. If such degrees of freedom are not in the menu, decoherence is not simply slow, it is essentially absent.

In effective field theory formulation of gravity dimensionless couplings are suppressed by negative powers of

---

* Corresponding author E-mail:
Dmitriy_Podolskiy@hms.harvard.edu

[1] Harvard Medical School, 77 Avenue Louis Pasteur, Boston, MA, 02115

[2] Wake Forest University, 1834 Wake Forest Rd., Winston-Salem, NC 27106

[1] There exist counter-arguments in favor of the existence of a genuine strong coupling limit for $d = 4$ [42].

he Planck mass M_P, which plays the role of UV cut-
•ff and becomes infinite in the decoupling limit $M_P \to$
∞. Decoherence times for arbitrary configurations of
quantum gravitational degrees of freedom also grow with
growing M_P although, as we shall see below, only loga-
ithmically slowly and become infinite at complete de-
oupling. If we recall that gravity is *almost* decoupled
rom physical matter in the real physical world, inef-
ectiveness of quantum gravitational decoherence does
not seem any longer so surprising. While matter de-
rees of freedom propagating on a fixed or slightly per-
urbed background space-time corresponding to a fixed
olution branch of the WdW equation decohere very
apidly, decoherence of different WKB solution branches
emains a question from the realm of quantum gravity.
hus, we would like to argue that in order to fit the in-
•ffectiveness of quantum gravitational decoherence and
. nearly perfectly decohered world which we experience
n experiments, some additional physical arguments are
ecessary based on properties of observer, in particular,
er/his ability to process and remember information.

This paper is organized as follows. We discuss de-
oherence in non-renormalizable quantum field theo-
ies and relation between non-renormalizable QFTs and
lassical statistical systems with first order phase tran-
tion in Section 2. We discuss decoherence in non-
enormalizable field theories in Section 3 using both
rst- and second-quantized formalisms. Section 4 is de-
oted to the discussion of decoherence in dS space-
me. We also argue that meta-observers in dS space-time
hould not be expected to experience effects of deco-
erence. Standard approaches to quantum gravitational
ecoherence based on analysis of WdW solutions and
naster equation for the density matrix of quantum grav-
ational degrees of freedom are reviewed in Section 5.
inally, we argue in Section 6 that one of the mecha-
isms responsible for the emergence of the arrow of time
 related to ability of observers to preserve information
bout experienced events.

Preliminary notes on non-renormalizable
field theories

o develop a quantitative approach for studying de-
oherence in non-renormalizable field theories, it is
astructive to use the duality between quantum field
heories in d space-time dimensions and statistical
hysics models in d spatial dimensions. In other words,
 gain some intuition regarding behavior of non-
enormalizable quantum field theories, one can first an-
lyze the behavior of their statistical physis counterparts

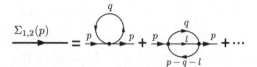

Figure 1   One- and two-loop contributions to $\Sigma(p)$ in $\lambda\phi^4$ EFT.

describing behavior of classical systems with appropriate symmetries near the phase transition.

Consider for example a large class of non-
renormalizable QFTs, which includes theories with
global discrete and continuous symmetries in the number of space-time dimensions higher than the upper
critical dimension d_{up}: $d > d_{up}$. Euclidean versions of
such theories are known to describe a vicinity of the
1st order phase transition on the lattice [27], and their
continuum limits do not formally exist[2]: even at close
proximity of the critical temperature $T = T_c$ physical
correlation length of the theory $\xi \sim m_{phys}^{-1} \sim (T - T_c)^{-1/2}$
never becomes infinite.

One notable example of such a theory is the $\lambda(\phi^2 - v^2)^2$ scalar statistical field theory, describing behavior of
the order parameter ϕ in the nearly critial system with
discrete Z_2 symmetry. This theory is trivial [25, 26] in
$d > d_{up} = 4$.[3] Triviality roughly follows from the observation that the effective dimensionless coupling falls off as
λ/ξ^{d-4}, when the continuum limit $\xi \to \infty$ is approached.

What does it mean physically? First, the behavior of
the theory in $d > 4$ is well approximated by mean field.
This can be readily seen when applying Ginzburg criterion for the applicability of mean field approximation
[28]: at $d > 4$ the mean field theory description is applicable arbitrarily close to the critical temperature. This is
also easy to check at the diagrammatic level: the two-
point function of the field ϕ has the following form in
momentum representation

$$\left\langle \phi(-p)\phi(p) \right\rangle \sim \left(p^2 + m_0^2 + \Sigma(p) \right)^{-1},$$

where $m_0^2 = a(T - T_c)$, and at one loop level (see Fig. 1)

$$\Sigma(p) \sim c_1 g \Lambda^2 + c_2 g \Lambda^2 \left(\frac{a(T - T_c)}{\Lambda^2} \right)^{d/2-1}, \tag{1}$$

[2] Similarly, Euclidean Z_2, $O(2)$ and $SU(N)$ gauge field theories all
known to possess a first order phase transition on the lattice at
$d > d_{up} = 4$.

[3] Most probably, it is trivial even in $d = 4$ [24], where it features a
Landau pole (although there exist arguments in favor of a non-
trivial behavior at strong coupling, see for example [42]).

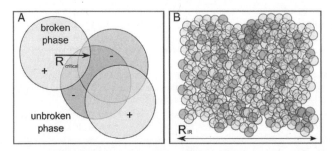

Figure 2 A possible configuration of order parameter in the Z_2 statistial model in $d \geq 5$ spatial dimensions. The left panel represents the configuration of the field at scales slightly larger than the critical radius $R_{\text{crit}} \sim \xi$, that coinsides with the size of bubbles of the true vacuum with broken Z_2 symmetry; $+$ corresponds to bubbles with the vacuum $+\Phi_0$ inside, and $- -$ to the bubbles with the vacuum $-\Phi_0$. At much larger scales of the order of R_{IR} given by the expression (2) $\langle \Phi \rangle = 0$ in average, as the contribution of multiple bubbles with $\Phi = +\Phi_0$ is compensated by the contribution of bubbles with $\Phi = -\Phi_0$.

where $g = \lambda \Lambda^{d-4}$ is the dimensionless coupling. The first term in the r.h.s. of (1) represents the mean field correction leading to the renormalization/redefinition of T_c. The second term is strongly suppressed at $d > 4$ in comparison to the first one. The same applies to any high order corrections in powers of λ as well as corrections from any other local terms $\sim \phi^6, \phi^8, \ldots, p^m \phi^n, \ldots$ in the effective Lagrangian of the theory.

As we see, the behavior of the theory is in fact simple despite its non-renormalizability; naively, since the coupling constant λ has a dimension $[l]^{d-4}$, one expects uncontrollable power-law corrections to observables and coupling constants of the theory. Nevertheless, as (1) implies, the perturbation theory series can be re-summed in such a way that only mean field terms survive. Physicswise, it is also clear why one comes to this conclusion. At $d > 4$ Z_2-invariant statistical physics models do not possess a second order phase transition, but of course do possess a first order one.[4] Behavior of the theory in the vicinity of the first order phase transition can always be described in the mean field approximation, in terms of the homogeneous order parameter $\Phi = \langle \phi \rangle$.

Our argument is not entirely complete as there is a minor culprit. Assume that an effective field theory with the EFT cutoff Λ coinciding with the physical cutoff is considered. Near the point of the 1st order phase transition, when the very small spatial scales (much smaller than the correlation length ξ of the theory) are probed, it is almost guaranteed that the probed physics is the one of the broken phase. The first order phase transition proceeds through the nucleation of bubbles of a critical size

$R \sim (T - T_c)^{-1/2} \sim \xi$, thus very small scales correspond to physics inside a bubble of the true vacuum $\langle \phi \rangle = \pm v$ and the EFT of the field $\delta \phi = \phi - \langle \phi \rangle$ is a good description of the behavior of the theory at such scales. As the spatial probe scale increases, such description will inevitably break down at the IR scale

$$R_{IR} \sim m^{-1} \exp \left(\frac{\text{Const.}}{\lambda m^{d-4}} \right)$$

$$\sim \frac{\Lambda^{\frac{d-4}{2}}}{\sqrt{g} v} \exp \left(\frac{\text{Const.} \Lambda^{(d-4)(d/2-1)}}{g^{d/2-1} v^{d-4}} \right), \tag{2}$$

where $m \sim \xi^{-1} \sim \sqrt{\lambda} v \to 0$ in the pre-critical limit. This scale is directly related to the nucleation rate of bubbles: at scales much larger than the bubble size R one has to take into account the stochastic background of the ensemble of bubbles of true vacuum on top of the false vacuum, and deviation of it from the the single-bubble background $\langle \phi \rangle = \pm v$ leads to the breakdown of the effective field theory description, see Fig. 2. Spatial homogeneity is also broken at scales $m^{-1} < l \ll R_{\text{IR}}$ by this stochastic background, and this large-scale spatial inhomogeneity is one of the reasons of the EFT description breakdown.

Finally, if the probe scale is much larger than $\xi \sim (T - T_c)^{-1/2}$ (say, roughly, of the order of R_{IR} or larger) the observer probes a false vacuum phase with $\langle \phi \rangle = 0$. Z_2 symmetry dictates the existence of two true minima $\langle \phi \rangle = \pm v$, and different bubbles have different vacua among the two realized inside them. If one waits long enough, the process of constant bubble nucleation will lead to self-averaging of the observed $\langle \phi \rangle$. As a result, the "true" $\langle \phi \rangle$ measured over very long spatial scales is always zero.

---

[4] This is equivalent to the statement that trvial theories do not admit continuum limit.

The main conclusion of this Section is that despite the EFT breakdown at both UV (momenta $p \gtrsim \Lambda$) and IR momenta $p \lesssim R_{IR}^{-1}$) scales, the non-renormalizable statistical $\lambda \phi^4$ theory perfectly remains under control: one can effectively use a description in terms of EFT at small scales $R_{IR}^{-1} \lesssim p \lesssim \Lambda$ and a mean field at large scales. In all cases, the physical system remains nearly completely described in terms of the homogeneous order parameter $\phi = \langle \phi \rangle$ or a "master field", as its fluctuations are almost decoupled. Let us now see what this conclusion means for the quantum counterparts of the discussed statistical physics systems.

Decoherence in relativistic non-renormalizable field theories

We first focus on the quantum field theory with global Z_2-symmetry. All of the above (possibility of EFT descriptions at both $R_{IR}^{-1} \lesssim E \lesssim \Lambda$ and $E \ll R_{IR}^{-1}$, breakdown of EFT at $E \sim \Lambda$ and $E \sim R_{IR}^{-1}$ with R_{IR} given by the expression (2)) can be applied to the quantum theory, but there is an important addition concerning decoherence, which we shall now discuss in more details.

1 Master field and fluctuations

As we discussed above, for the partition function of the Z_2-invariant statistical field theory describing a vicinity of a first order phase transition $\frac{T - T_c}{T_c} \ll 1$ one approximately has

$$Z = \int \mathcal{D}\phi \exp\left(-\int d^d x \left(\frac{1}{2}(\partial\phi)^2 \pm \frac{1}{2}m^2\phi^2 + \frac{1}{4}\lambda\phi^4 + \dots\right)\right) \approx \tag{3}$$

$$\approx \int d\Phi \exp\left(\mp\frac{1}{2}V_d m^2\Phi^2 - \frac{1}{4}V_d\lambda\Phi^4 - V_d\mu\Phi\right), \tag{4}$$

where V_d is the d–volume of the system, and $d \geq 5$ as in the previous Section. Physically, the spatial fluctuations of the order parameter ϕ are suppressed, and the system is well described by statistical properties of the homogeneous order parameter $\Phi \sim \langle \phi \rangle$.

The Wick rotated quantum counterpart of the statistical physics model (3) is determined by the expression for the quantum mechanical "amplitude"

$$A(\Phi_0, t_0; \Phi, t) \approx \int d\Phi \exp\left(iV_{d-1}T\left(\mp\frac{1}{2}m^2\Phi^2 - \frac{1}{4}\lambda\Phi^4\right)\right) = $$

$$\int_{\Phi(t_0)=\Phi_0}^{\Phi(t)=\Phi} \mathcal{D}\Phi \exp\left(iV_{d-1}\int_{t_0}^t dt\left(\mp\frac{1}{2}m^2\Phi^2 - \frac{1}{4}\lambda\Phi^4\right)\right), \tag{5}$$

written entirely in terms of the "master field" Φ (as usual, $V_{d-1} = \int d^{d-1}x$ is the volume of $(d-1)$-dimensional space). In other words, in the first approximation the non-renormalizable $\lambda\phi^4$ theory in $d \geq 5$ dimensions can be described in terms of a master field Φ, roughly homogeneous in space-time. As usual, the wave function of the field can be described as

$$\Psi(\Phi, t) \sim A(\Phi_0, t_0; \Phi, t),$$

where Φ_0 and t_0 are fixed, while Φ and t are varied, and the density matrix is given by

$$\rho(\Phi, \Phi', t) = \text{Tr}\Psi(\Phi, t)\Psi^*(\Phi', t), \tag{6}$$

where the trace is taken over the degrees of freedom not included into Φ and Φ', namely, fluctuations of the field $\delta\phi$ above the master field configuration Φ. The contribution of the latter can be described using the prescription

$$A \sim \int d\Phi \mathcal{D}\delta\phi \exp\left(iV_{d-1}T\left(\mp\frac{1}{2}m^2\Phi^2 - \frac{1}{4}\lambda\Phi^4\right)\right) \times$$

$$\times \exp\left(i\int d^d x\left(\frac{1}{2}(\partial\delta\phi)^2 \mp \frac{1}{2}m^2\delta\phi^2 - \frac{3}{2}\lambda\Phi^2\delta\phi^2\right.\right.$$

$$\left.\left. -\lambda\Phi^3\delta\phi - \lambda\Phi\delta\phi^3 - \dots\right)\right). \tag{7}$$

In the "mean field" approximation (corresponding to the continuum limit) $\lambda \to 0$ fluctuations $\delta\phi$ are completely decoupled from the master field Φ, making (5) a good approximation of the theory. To conclude, one physical consequence of the triviality of statistical physics models describing vicinity of a first order phase transition is that in their quantum counterparts decoherence of entangled states of the master field Φ does not proceed.

3.2 Decoherence in the EFT picture

When the correlation length $\xi \sim m_{\text{phys}}^{-1}$ is large but finite, decoherence takes a finite but large amount of time, essentially, as we shall see, determined by the magnitude of ξ. This time scale will now be estimated by two different methods.

As non-renormalizable QFTs admit an EFT description (which eventually breaks down), dynamics of decoherence in such theories strongly depends on the probe scale, coarse-graining effectively performed by the observer. Consider a spatio-temporal coarse-graining scale $l > \Lambda^{-1}$ and assume that all modes of the field ϕ with energies/momenta $l^{-1} < p \ll \Lambda$ represent the "environment", and interaction with them leads to the decoherence of the observed modes with momenta $p < l^{-1}$.

If also $p > R_{IR}^{-1}$, EFT expansion near $\langle\phi\rangle$ is applicable. In practice, similar to Kenneth Wilson's prescription for renormalization group analysis, we separate the field ϕ into the fast, ϕ_f, and slow, ϕ_s, components, considering ϕ_f as an environment, and since translational invariance holds "at large", ϕ_s and ϕ_f are linearly separable.[5]

The density matrix $\rho(t, \phi_s, \phi_s')$ of the "slow" field or master field configurations is related to the Feynman-Vernon influence functional $S_I[\phi_1, \phi_2]$ of the theory [21] according to

$$\rho(t, \phi_s, \phi_s') = \int d\phi_0 d\phi_0' \rho(t, \phi_0, \phi_0') \times$$

$$\times \int_{\phi_0}^{\phi_s} d\phi_1 \int_{\phi_0'}^{\phi_s'} d\phi_2 \exp\left(iS[\phi_1] - iS[\phi_2]\right.$$
$$\left. + iS_I[\phi_1, \phi_2]\right), \tag{8}$$

where

$$S[\phi_{1,2}] = \int d^d x \left(\frac{1}{2}(\partial\phi_{1,2})^2 - \frac{1}{2}m^2\phi_{1,2}^2 - \frac{1}{4}\lambda\phi_{1,2}^4\right), \tag{9}$$

and

$$S_I = -\frac{3}{2}\lambda \int d^d x \Delta_F(x, x)(\phi_1^2 - \phi_2^2) +$$

$$+ \frac{9\lambda^2 i}{4} \int d^d x\, d^d y \phi_1^2(x)(\Delta_F(x, y))^2 \phi_1^2(y) -$$

$$- \frac{9\lambda^2 i}{2} \int d^d x\, d^d y \phi_1^2(x)(\Delta_-(x, y))^2 \phi_2^2(y) +$$

$$\frac{9\lambda^2 i}{4} \int d^d x\, d^d y \phi_2^2(x)(\Delta_D(x, y))^2 \phi_2^2(y) + \dots,$$

where $\phi_{1,2}$ are the Schwinger-Keldysh components of the field ϕ_s, and $\Delta_{F,-,D}$ are Feynman, negative frequency Wightman and Dyson propagators of the "fast" field ϕ_f respectively.[6] It is easy to see that the expression (9) is essentially the same as (7), that is of no surprise since an observer with an IR cutoff cannot distinguish between ϕ and ϕ_s.

The part of the Feynman-Vernon functional (10) that is interesting for us can be rewritten as

$$S_I = i\lambda^2 \int d^d x\, d^d y (\phi_1^2(x) - \phi_2^2(x))\nu(x - y)(\phi_1^2(y) - \phi_2^2(y)) -$$
$$- \lambda^2 \int d^{d+1} x\, d^{d+1} y (\phi_1^2(x) - \phi_2^2(x))\mu(x - y) \tag{11}$$
$$\times \left(\phi_1^2(y) + \phi_2^2(y)\right) + \dots$$

(note that non-trivial effects including the one of decoherence appear in the earliest only at the second order in λ).

An important observation to make is that since the considered non-renormalizable theory becomes trivial in the continuum limit, see (5), the kernels μ and ν can be approximated as local, i.e., $\mu(x - y) \approx \mu_0\delta(x - y)$, $\nu(x - y) \approx \nu_0\delta(x - y)$. This is due to the fact that fluctuations $\delta\phi \sim \phi_f$ are (almost) decoupled from the master field $\Phi \sim \phi_s$ in the continuum limit, their contribution to (9) is described by the (almost) *Gaussian* functional. Correspondingly, if one assumes factorization and Gaussianity of the initial conditions for the modes of the "fast" field ϕ_f, the Markovian approximation is valid for the functional (9), (10).

A rather involved calculation (see [21]) then shows that the density matrix (8) is subject to the master equation

$$\frac{\partial\rho(t, \phi_s, \phi_s')}{\partial t} = -\int d^{d-1} x [H_I(x, \tau), \rho] + \dots, \tag{12}$$

$$H_I \approx \frac{1}{2}\lambda^2\nu_0(\phi_s^2(\tau, x) - \phi_s'^2(\tau, x))^2,$$

---

[5] A note should be taken at this point regarding the momentum representation of the modes. As usual, ϕ_f is defined as integral over Fourier modes of the field with small momenta. As explained above, the quantum theory with existing continuum limit is a Wick-rotated counterpart of the statistical physics model describing a second order phase transition. In the vicinity of a second order phase transition broken and unbroken symmetry phases are continuously intermixed together, which leads to the translational invariance of correlation functions of the order parameter ϕ. In the case of the first order phase transition, such invariance is strictly speaking broken in the presence of stochastic background of nucleating bubbles of the broken symmetry phase, see the discussion in the previous Section. Therefore, the problem "at large" rewritten in terms of ϕ_f and ϕ_s becomes of Caldeira-Legett type [43]. If we focus our attention on the physics at scales smaller than the bubble size, translational invariance does approximately hold, and we can consider ϕ_s and ϕ_f as linearly separable (if they are not, we simply diagonalize the part of the Hamiltonian quadratic in ϕ).

[6] Here, we kept only the leading terms in $\lambda \sim \xi^{4-d}$ as higher loops as well as other non-renormalizable interactions provide contributions to the FV functional, which are subdominant (and vanishing in the continuum limit $\xi \to \infty$).

where only terms of the Hamiltonian density H_I, which lead to the exponential decay of non-diagonal matrix elements of ρ are kept explicitly, while ... denote oscillatory terms.

The decoherence time can easily be estimated as follows. If only "quasi"-homogeneous master field is kept in (12), the density matrix is subject to the equation

$$\frac{\partial \rho(t, \Phi, \Phi')}{\partial t} = -\frac{1}{2}\lambda^2 v_0 V_{d-1}[(\Phi - \Phi')^2(\Phi + \Phi')^2, \rho]$$

$$= -\frac{1}{2}\lambda^2 v_0 V_{d-1}[(\Phi - \Phi')^2 \bar{\Phi}^2, \rho], \quad (13)$$

where $\bar{\Phi} = \frac{1}{2}(\Phi + \Phi')$. We expect that $\bar{\Phi}$ is close to (but does not necessarily coincides with) the minimum of the potential $V(\Phi)$, which will be denoted Φ_0 in what follows. For $\Phi \approx \Phi'$, i.e., diagonal matrix elements of the density matrix the decoherence effects are strongly suppressed. For the matrix elements with $\Phi \neq \Phi'$ the decoherence rate is determined by

$$\Gamma = \frac{1}{2}\lambda^2 v_0 V_{d-1}(\Phi - \Phi')^2\bar{\Phi}^2 \approx \frac{1}{2}\lambda^2 v_0 V_{d-1}(\Phi - \Phi')^2\Phi_0^2. \quad (14)$$

Thus, the decoherence time scale in this regime is

$$t_D \sim \frac{1}{\lambda^2 v_0 V_{d-1}(\Phi - \Phi')^2\Phi_0^2}. \quad (15)$$

It is possible to further simplify this expression. First of all, one notes that λ_{renorm} will be entering the final answer instead of the bare coupling λ. As was discussed above (and shown in details in [25, 26]), the dimensionless renormalized coupling g_{renorm} is suppressed in the continuum limit as $\frac{\text{Const}}{\xi^{d-4}}$, where ξ is the physical correlation length. Second, the physical volume V satisfies the relation $V \lesssim \xi^{d-1}$ (amounting to the statement that the continuum limit corresponds to correlation length being of the order of the system size). Finally, $\Phi_0^2 \sim \frac{m_{\text{ren}}^2}{\lambda} \sim \xi^{d-6}$, i.e., every quantity in (15) can be presented in terms of the physical correlation length ξ only. This should not be surprising. As was argued in the previous sections, the mean field theory description holds effectively in the limit $\Lambda \to \infty$ (or $\xi \to \infty$), which is characterized by uncoupling of fluctuations from the mean field Φ. Self-coupling of fluctuations $\delta\phi$ is also suppressed in the same limit, thus the physical correlation length becomes a single parameter defining the theory. The only effect of taking into account next orders in powers of λ (or other interactions!) in the effective action (9) and the Feynman-Vernon functional (10) is the redefinition of ξ, which ultimately has to be determined

from observations. In this sense, (15) holds to all orders in λ, and it can be expected that

$$t_D \gtrsim \text{Const}.\xi \cdot (\xi/\delta\xi), \quad (16)$$

where $\delta\xi \sim |\Phi - \Phi'|$ universally for all Φ, Φ' of physical interest.

According to the expressions (15), (16) decay of non-diagonal elements of the density matrix $\rho(t, \Phi_1, \Phi_2)$ would take much longer than ξ/c (where c is the speed of light) for $|\Phi_1 - \Phi_2| \ll |\Phi_1 + \Phi_2|$. It still takes about $\sim \xi/c$ for matrix elements with $|\Phi_1 - \Phi_2| \sim |\Phi_1 + \Phi_2|$ to decay, a very long time in the limit $\xi \to \infty$.

Finally, if $\bar{\Phi} \neq \Phi_0$, i.e., the "vacuum" is excited, $\bar{\Phi}$ returns to minimum after a certain time and fluctuates near it. It was shown in [21] that the field Φ is subject to the Langevin equation

$$2\mu_0\Phi_0\frac{d\Phi}{dt} + m^2(\Phi - \Phi_0) \approx \Phi_0\xi(t), \quad (17)$$

$$\langle \xi(t) \rangle = 0,$$

$$\langle \xi(t)\xi(t') \rangle = v_0\delta(t - t'),$$

where the random force is due to the interaction between the master field Φ and the fast modes $\delta\phi$, determined by the term $\frac{3}{2}\lambda\Phi^2\delta\phi^2$ in the effective action. (The Eq. (17) was derived by application of Hubbard-Stratonovich transformation to the effective action for the fields Φ and $\delta\phi$ and assuming that Φ is close to Φ_0.) The average

$$\langle \Phi \rangle - \Phi_0 \approx (\Phi_{\text{init}} - \Phi_0)\exp\left(-\frac{m^2}{2\mu_0\Phi_0}(t - t_{\text{init}})\right),$$

so the master field rolling towards the minimum of its potential plays a role of "time" in the theory. The roll towards the minimum Φ_0 is very slow, as the rolling time $\sim \frac{\mu_0\Phi_0}{m^2} \sim \frac{\mu_0}{\sqrt{\lambda}m} \sim \xi^{d-3}$ is large in the continuum limit $\xi \to 0$. Once the field reaches the minimum, there is no "time", as the master field Φ providing the function of a clock is minimized. The decoherence would naively be completely absent for the superposition state of vacua $\pm\Phi_0$ as follows from (14). However, the physical vacuum as seen by a coarse-grained observer is subject to the Langevin equation (17) even in the closest vicinity of $\Phi = \pm\Phi_0$, and the fluctuations $\langle (\Phi - \Phi_0)^2 \rangle$ are never zero; one roughly has

$$\langle (\Phi - \Phi_0)^2 \rangle \sim \frac{\Phi_0 v_0}{m\mu_0},$$

which should be substituted in the estimate (15) for matrix elements with $\Phi \approx \Phi' \approx \Phi_0$.

What was discussed above holds for coarse-graining scales $p > R_{IR}^{-1}$, where R_{IR} is given by the expression (2). If the coarse-graining scale is $p \lesssim R_{IR}^{-1}$, the EFT description breaks down, since at this scale the effective dimensionless coupling between different modes becomes of the order 1, and the modes contributing to ϕ_s and ϕ_f can no longer be considered weakly interacting. However, we recall that at probe scales $l > R_{IR}$ the unbroken phase mean field description is perfectly applicable (see above). This again implies extremely long decoherence time scales.

The emergent physical picture is the one of entangled states with coherence surviving during a very long time (at least $\sim \xi/c$) on spatial scales of the order of at least ξ. The largeness of the correlation length ξ in statistical physics models describing the vicinity of a first order phase transition implied a large scale correlation at the spatial scales $\sim \xi$. As was suggested above, the decoherence is indeed very ineffective in such theories. We shall see below that the physical picture presented here has a very large number of analogies in the case of decoherence in quantum gravity.

3.3 Decoherence in functional Schrodinger picture

Let us now perform a first quantization analysis of the theory and see how decoherence emerges in this analysis. As the master field Φ is constant in space-time, the field state approximately satisfies the Schrodinger equation

$$\hat{H}_\Phi |\Psi(\Phi)\rangle = E_0 |\Psi(\Phi)\rangle,$$

where the form of the Hamiltonian $\hat{H}_\Phi$ follows straightforwardly from (5):

$$\hat{H}_\Phi = -\frac{1}{2} V_{d-1} \frac{\partial^2}{\partial \Phi^2} \pm \frac{1}{2} V_{d-1} m^2 \Phi^2 + \frac{1}{4} V_{d-1} \lambda \Phi^4.$$

The physical meaning of E_0 is the vacuum energy of the scalar field, which one can safely choose to be 0.

Next, one looks for the quasi-classical solution of the Schrodinger equation of the form $\Psi_0(\Phi) \sim \exp(iS_0(\Phi))$. The wave function of fluctuations $\delta\phi$ (or ϕ_f using terminology of the previous Subsection) in turn satisfies the Schrodinger equation

$$i \frac{\partial \psi(\Phi, \phi_f)}{\partial \tau} = \hat{H}_{\delta\phi} \psi(\Phi, \phi_f), \tag{18}$$

where $\frac{\partial}{\partial \tau} = \frac{\partial S_0}{\partial \Phi} \frac{\partial}{\partial \Phi}$ and $\hat{H}_{\delta\phi}$ is the Hamiltonian of fluctuations $\delta\phi$,

$$\hat{H}_{\delta\phi} = \int d^{d-1}x \left(-\frac{1}{2} \frac{\partial^2}{\partial \delta\phi^2} + \frac{1}{2} (\nabla \delta\phi)^2 + V(\Phi, \delta\phi) \right),$$

where $V(\Phi, \delta\phi) = \pm \frac{1}{2} m^2 \delta\phi^2 + \frac{3}{2} \lambda \Phi^2 \phi^2$, and the full state of the field is $\Psi(\Phi, \delta\phi) \sim \Psi_0(\Phi) \psi(\Phi, \delta\phi) \sim \exp(iS_0(\Phi)) \psi(\Phi, \delta\phi)$ (again, we naturally assume that the initial state was a factorized Gaussian). It was previously shown (see [17] and references therein) that the "time"-like affine parameter τ in (18) coincides in fact with the physical time t.

Writing down the expression for the density matrix of the master field Φ

$$\rho(t, \Phi_1, \Phi_2) = \mathrm{Tr}_{\delta\phi}(\Psi(\Phi_1, \delta\phi) \Psi^*(\Phi_2, \delta\phi))$$

$$= \rho_0 \int \mathcal{D}\delta\phi \, \psi(\tau, \Phi_1, \delta\phi) \psi^*(\tau, \Phi_2, \delta\phi), \tag{19}$$

where

$$\rho_0 = \exp(iS_0(\Phi_1) - iS_0(\Phi_2)),$$

$$S_0(\Phi) = \frac{1}{2} V_{d-1}(\dot{\Phi})^2 \mp \frac{1}{2} V_{d-1} m^2 \Phi^2 - \frac{1}{4} V_{d-1} \lambda \Phi^4,$$

one can then repeat the analysis of [17]. Namely, one takes a Gaussian ansatz for $\psi(\tau, \Phi, \delta\phi)$ (again, this is validated by the triviality of the theory)

$$\psi(\tau, \Phi, \delta\phi) = N(\tau) \exp \left(-\int d^{d-1} p \, \delta\phi(p) \Omega(p, \tau) \delta\phi(p) \right),$$

where N and Ω satisfy the equations

$$i \frac{d \log N(\tau)}{d\tau} = \mathrm{Tr}\Omega, \tag{20}$$

$$-i \frac{\partial \Omega(p, \tau)}{\partial \tau} = -\Omega^2(p, \tau) + \omega^2(p, \tau), \tag{21}$$

$$\omega^2(p, \tau) = p^2 + m^2 + 3\lambda \Phi^2 + \ldots,$$

and the trace denotes integration over modes with different momenta:

$$\mathrm{Tr}\Omega = V_{d-1} \int \frac{d^{d-1}p}{(2\pi)^{d-1}} \Omega(p, \tau).$$

The expression for $N(t)$ can immediately be found using the Eq. (20) and the normalization condition

$$\int \mathcal{D}\delta\phi |\psi(\tau, \Phi, \delta\phi)|^2 = 1$$

(if $N(\tau) = |N(\tau)| \exp(i\xi(\tau))$, the former completely determines the absolute value $|N(\tau)|$, while the latter — the phase $\xi(\tau)$). Then, after taking the Gaussian functional integration in (19), the density matrix can be rewritten in terms of the real part of $\Omega(p, \tau)$ as

$$\rho(\Phi_1, \Phi_2) \approx \frac{\rho_0 \sqrt{\det(\text{Re}(\Phi_1)) \det(\text{Re}(\Phi_2))}}{\sqrt{\det(\Omega(\Phi_1) + \Omega^*(\Phi_2))}}$$

$$\times \exp\left(-i \int^t dt' \cdot (\text{Re}\Omega(\Phi_1) - \text{Re}\Omega(\Phi_2))\right).$$

Assuming the closeness of Φ_1 and Φ_2 and following [17] we expand

$$\Omega(\Phi_2) \approx \Omega(\bar{\Phi}) + \Omega'(\bar{\Phi})\Delta + \frac{1}{2}\Omega''(\bar{\Phi})\Delta^2 + \ldots,$$

where again $\bar{\Phi} = \frac{1}{2}(\Phi_1 + \Phi_2)$, $\Delta = \frac{1}{2}(\Phi_1 - \Phi_2)$, and keep terms proportional to Δ^2 only. A straightforward but lengthy calculation shows that the exponentially decaying term in the density matrix has the form

$$\exp(-D) = \exp\left(-\text{Tr}\frac{|\Omega'(\bar{\Phi})|^2}{(\text{Re}\Omega(\bar{\Phi}))^2}\Delta^2\right), \tag{22}$$

where D is the decoherence factor, and the decoherence time can be directly extracted from this expression. To do so, we note that $\Omega(\Phi)$ is subject to the Eq. (21). When $\bar{\Phi} = \Phi_0$, one has $\Omega^2 = \omega^2$, and Ω does not have any dynamics according to (21). However, if $\Phi_{1,2} \neq \bar{\Phi}_0$, $\Omega^2 \neq \omega^2$. As the dynamics of Φ is slow (see Eq. (17)), one can consider ω as a function of the constant field Φ and integrate the Eq. (21) directly. As the time t enters the solution of this equation only in combination ωt, one immediately sees that the factor (22) contains a term $\sim t$ in the exponent, defining the decoherence time. The latter coincides with the expression (15) derived in the previous Section as should have been expected.

Thus, the main conclusion of this Section is that the characteristic decoherence time scale in non-renormalizable field theories akin to the $\lambda\phi^4$ theory in number of dimensions higher than 4 is at least of the order of the physical correlation length ξ of the theory, which is taken to be large in the continuum limit. Thus, decoherence in the nearly continuum limit is very ineffective for such theories.

4 Decoherence of QFTs on curved space-times

Before proceeding to the discussion of the case of gravity, it is instructive to consider how the dynamics of decoherence of a QFT changes once the theory is set on a curved space-time. As we shall see in a moment, even when the theory is renormalizable (the number of space-time dimensions $d = d_{\text{up}}$), the setup features many similarities with the case of a non-renormalizable field theory in the flat space-time discussed in the previous Section.

Consider a scalar QFT with potential $V(\phi) = \frac{1}{2}m^2\phi^2 + \frac{1}{4}\lambda\phi^4$ in 4 curved space-time dimensions. Again, we assume the nearly critical $"T \to T_c"$ case, and that is why the renormalized quadratic term $\frac{1}{2}m^2\phi^2$ determining the correlation length of the theory $\xi \sim m_{\text{renorm}}^{-1}$ is set to vanish (compared to the cutoff scale Λ, again for definiteness $\Lambda \sim M_P$).

The scale ξ is no longer the only relevant one in the theory. The structure of the Riemann tensor of the space-time (the latter is assumed to be not too curved) introduces new infrared scales for the theory, and the dynamics of decoherence in the theory depends on relation between these scales and the mass scale m. Without a much loss of generality and for the sake of simplicity, one can consider a dS_4 space-time characterized by a single such scale (cosmological constant) related to the Ricci curvature of the background space-time. It is convenient to write

$$V(\phi) = V_0 + \frac{1}{2}m^2\phi^2 + \frac{1}{4}\lambda\phi^4,$$

assuming that the V_0 term dominates in the energy density.

At spatio-temporal probe scales much smaller than the horizon size $H_0^{-1} \sim \frac{M_p}{\sqrt{V_0}}$ one can choose the state of the field to be the Bunch-Davies (or Allen-Mottola) vacuum or an arbitrary state from the same Fock space. Procedures of renormalization, construction the effective action of the theory and its Feynman-Vernon influence function are similar to the ones for QFT in Minkowski space-time. Thus, so is the dynamics of decoherence due to tracing out unobservable UV modes; the decoherence time scale is again of the order of the physical correlation length of the theory:

$$t_D \sim \xi \sim m_{\text{renorm}}^{-1},$$

in complete analogy with the estimate (16). This standard answer is replaced by

$$t_D \sim H_0^{-1}, \tag{23}$$

when the mass of the field becomes smaller than the Hubble scale, $m^2 \ll H_0^2$, and the naive correlation length ξ exceeds the horizon size of dS_4. (The answer (23) is correct up to a logarithmic prefactor $\sim \log(H_0)$.)

It is interesting to analyze the case $m^2 \ll H_0^2$ in more details. The answer (4) is only applicable for a physical observer living inside a single Hubble patch. How does the decoherence of the field ϕ look like from the point of view of *a meta-observer*, who is able to probe the super-horizon large scale structure of the field ϕ?[7] It is well-known [29, 30] that the field ϕ in the planar patch of dS_4 coarse-grained at the spatio-temporal scale of cosmological horizon H_0^{-1} is (approximately) subject to the Langevin equation

$$3 H_0 \frac{d\phi}{dt} = -m^2\phi - \lambda\phi^3 + f(t),$$ (24)

$$\left\langle f(t) f(t') \right\rangle = \frac{3 H_0^4}{4\pi^2} \delta(t - t'),$$

where average is taken over the Bunch-Davies vacuum, very similar to (17), but with the difference that the amplitude of the white noise and the dissipation coefficient are correlated with each other. The corresponding Fokker-Planck equation

$$\frac{\partial P(t, \phi)}{\partial t} = \frac{1}{3 H_0} \frac{\partial}{\partial \phi} \left(\frac{\partial V}{\partial \phi} P(t, \phi) \right) + \frac{H_0^3}{8\pi^2} \frac{\partial^2 P}{\partial \phi^2}$$ (25)

describes behavior of the probability $P(t, \phi)$ to measure a given value of the field ϕ at a given moment of time at a given point of coarse-grained space. Its solution is normalizable and has an asymptotic behavior

$$P(t \to \infty, \phi) \sim \frac{1}{V(\phi)} \exp\left(-\frac{8\pi^2 V(\phi)}{3 H_0^4} \right)$$ (26)

As correlation functions of the coarse-grained field ϕ are calculated according to the prescription

$$\langle \phi^n(t, x) \rangle \sim \int d\phi \cdot \phi^n P(\phi, t),$$

(note that two-, three, etc. point functions of ϕ are zero, and only one-point correlation functions are non-trivial) what we are dealing with in the case (26) is nothing but

---

7   This question is not completely meaningless, since a setup is possible in which the value of V_0 suddenly jumps to zero, so that the background space-time becomes Minkowski in the limit $M_P \to 0$, and the field structure inside a single Minkowski lightcone becomes accessible for an observer. If her probe/coarse-graining scale is $l > H_0^{-1}$, this is the question which we are trying to address.

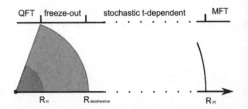

Figure 3 The hierarchy of decoherence scales for a metaobserver in dS_D space. $R_H \sim H_0^{-1}$ represents the Hubble radius, at comoving scales $< H_0^{-1}$ the correct physical description is the one in terms of interacting QFT in a de Sitter-invariant vacuum state; the freeze-out of modes leaving the horizon, vanishing of the decaying mode and decoherence of the background field ("master field" Φ) proceeds at comoving scales $R_H < l < R_{\text{decoherence}}$, where the latter is by a few efoldings larger than the former, see the next Section; at $R_H < l < R_{\text{decoherence}}$ the field Φ and related observables are subject to the Langevin equation (24) and represent a stochastic time-dependent background of Hubble patches; at comoving scales $> R_{\text{IR}}$ given by (27) the stochastic field Φ reaches the equilibrium solution (26) of the Fokker-Planck equation (25), and the notion of time is not well defined; the correct description of the theory is in terms of the mean/master field with partition function given by (26).

a mean field theory with a free energy $F = \frac{8\pi^2}{3} V(\phi)$ calculated as an integral of the mean field ϕ over the 4−volume $\sim H_0^{-4}$ of a single Hubble patch. As we have discussed in the previous Section, decoherence is not experienced as a physical phenomenon by the meta-observer at all. In fact, the coarse-graining comoving scale l_c separating the two distinctly different regimes of a weakly coupled theory with a relatively slow decoherence and a mean field theory with entirely absent decoherence is of the order

$$R_{\text{IR}} \sim H_0^{-1} \exp(S_{dS}),$$ (27)

where $S_{dS} = \frac{\pi M_P^2}{H_0^2}$ is the de Sitter entropy (compare this expression with (2)).

Overall, the physical picture which emerges for the scalar quantum field theory on dS_4 background is not very different from the one realized for the non-renormalizable $\lambda\phi^4$ field theory in Minkowski space-time, see Fig. 3:

- for observers with small coarse-graining (comoving) scale $l < H_0^{-1}$ the decoherence time scale is at most H_0^{-1}, which is rather large physically (of the order of cosmological horizon size for a given Hubble patch),
- for a meta-observer with a coarse-graining (comoving) scale $l > R_{\text{IR}}$, where R_{IR} is given by

(27), the decoherence is absent entirely, and the underlying theory is experienced as a mean field by such meta-observers.

Another feature of the present setup which is consistent with the behavior of a non-renormalizable field theory in a flat space-time is the breakdown of the effective field theory for the curvature perturbation in the IR [31] (as well as IR breakdown of the perturbation theory on a fixed dS_4 background) [32], compare with the discussion in Section 3. The control on the theory can be recovered if the behavior of observables in the EFT regime is glued to the IR mean field regime of eternal inflation [33].

5 Decoherence in quantum gravity

Given the discussions of the previous two Sections, we are finally ready to muse on the subject of decoherence in quantum gravity, emergence of time and the cosmological arrow of time, focusing on the case of $d = 3 + 1$ dimensions. The key observation for us is that the *critical number of dimensions for gravity is* $d_{up} = 2$, thus it is tempting to hypothesize that the case of gravity might have some similarities with the non-renormalizable theories discussed in Section 3.

One can perform the analysis of decoherence of quantum gravity following the strategy represented in Section 3.2, i.e., studying EFT of the second-quantized gravitational degrees of freedom, constructing the Feynman-Vernon functional for them and extracting the characteristic decoherence scales from it (see for example [34]). However, it is more convenient to follow the strategy outlined in Section 3.3. Namely, we would like to apply the Born-Oppenheimer approximation [17] to the Wheeler-de Witt equation

$$\hat{q}\Psi = \left(\frac{16\pi G_{ijkl}}{M_P^2} \frac{\partial^2}{\partial h_{ij} \partial h_{kl}} + \sqrt{h^{(3)}} M_P^2 R - \hat{H}_m \right) \Psi = 0 \tag{28}$$

describing behavior of the relevant degrees of freedom gravity + a free massive scalar field with mass m and the Hamiltonian $\hat{H}_m$). As usual, gravitational degrees of freedom include functional variables of the ADM split: scale factor a, shift and lapse functions N_μ and the transverse traceless tensor perturbations h_{ij}. The WdW equation (28) does not contain time at all; similar to the case of the Fokker-Planck equation (25) for inflation [29] the scale factor a replaces it. Time emerges only after a particular WKB branch of the solution Ψ is picked, and the WKB piece $\psi(a) \sim \exp(iS_0)$ of the wave function Ψ

is explicitly separated from the wave functions of the multipoles ψ_n [17], so that the full state is factorized: $\Psi = \psi(a) \prod_n \psi_n$. Similar to the case discussed in Section 3.3, the latter then satisfy the functional Schrodinger equations

$$i\frac{\partial \psi_n}{\partial \tau} = \hat{H}_n \psi_n, \tag{29}$$

(compare to (18)). In other words, as gravity propagates in $d = 4 > d_{up} = 2$ space-time dimensions, we assume a almost complete decoupling of the multipoles ψ_n from each other. Their Hamiltonian $\hat{H}_n$ is expected to be Gaussian with possible dependence on a: ψ_n's are analogous to the states ψ described by (18) in the case of a non-renormalizable field theory in the flat space-time. (We note though that this assumption of ψ_n decoupling might, generally speaking, break down in the vicinity of horizons such as black hole horizons, where the effective dimensionality of space-time approaches 2, the critical number of dimensions for gravity.)

The affine parameter τ along the WKB trajectory is again defined according to the prescription

$$\frac{\partial}{\partial t} = \frac{\partial}{\partial a} S_0 \frac{\partial}{\partial a}$$

and starts to play a role of physical time [17]. One is motivated to conclude that the emergence of time is related to the decoherence between different WKB branches of the WdW wave function Ψ, and such emergence can be quantitatively analyzed.

It was found in [17] by explicit calculation that the density matrix for the scale factor a behaves as

$$\rho(a_1, a_2) \sim \exp(-D)$$

with the decoherence factor for a single WKB branch of the WdW solution is given by

$$D \sim \frac{m^3}{M_P^3}(a_1 + a_2)(a_1 - a_2). \tag{30}$$

We note the analogy of this expression with the expression (22) derived in the the Section 3.3: decoherence vanishes in the limit $a_1 = a_2$ (or $a_1 = -a_2$) and is suppressed by powers of cutoff M_P (m/M_P can roughly be considered as a dimensionless effective coupling between matter and gravity). In particular, decoherence is completely absent in the decoupling limit $M_P \to \infty$.

To estimate the involved time scales, let us consider for definiteness the planar patch of dS_4 with $a(t) \sim \exp(H_0 t)$. It immediately follows from (30) that the single WKB branch decoherence only becomes effective after

$$H_0 t_d \gtrsim \log \left(\frac{M_P^3}{m^3(a_1 - a_2)} \right) \tag{31}$$

Hubble times, a logarithmically large number of efoldings in the regime of physical interest, when $M_P \gg m_{\text{phys}} \to 0$ (see also discussion of the decoherence of cosmic fluctuations in [35], where a similar logarithmic amplification with respect to a single Hubble time is found). Similarly, the decoherence scale between the two WKB branches of the WdW solution (corresponding to expansion and contraction of the inflating space-time)

$$\psi \sim c_1 e^{iS_0} + c_2 e^{-iS_0}$$

can be shown to be somewhat smaller [17, 34]: one finds for the decoherence factor

$$D \sim \frac{m H_0^2 a^3}{M_P^3},$$

and the decoherence time (derived from the bound $D(t_d) \gtrsim 1$) is given by

$$H_0 t_d \gtrsim \log \left(\frac{M_P^3}{m H_0^2} \right), \tag{32}$$

still representing a logarithmically large number of efoldings. Taking for example $m \sim 100$ GeV and $H_0 \sim 10^{-42}$ GeV one finds $H_0 t_d \sim 300$. Even for inflaitonary energy scale $H_0 \sim 10^{16}$ GeV the decoherence time scale is given by $H_0 t_d \sim 3$ inflationary efoldings, still a noticeable number. Interestingly, it also takes a few efoldings for the modes leaving the horizon to freeze and become quasiclassical.

Note that (a) H_0 does not enter the expression (31) at all, and it can be expected to hold for other (relatively spatially homogeneous) backgrounds beyond dS_d, (b) (31) is proportional to powers of effective dimensionless coupling between matter and gravity, which gets suppressed in the "continuum"/decoupling limit by powers of cutoff, (c) decoherence is absent for the elements of the density matrix with $a_1 = \pm a_2$. These analogies allow us to expect that a set of conclusions similar to the ones presented in Sections 3 and 23 would hold for gravity on other backgrounds as well:

• we expect the effective field theory description of gravity to break down in the IR at scales $l \sim l_{\text{IR}}$[8]; the

---

[8] Space-like interval $l = \int ds$ connecting two causally unconnected events.

latter is exponentially larger than the characteristic scale of curvature radius $\sim R_H$ of the background; we roughly expect

$$l_{\text{IR}} \sim R_H \exp \left(\frac{\text{Const.}}{M_P R_H} \right), \tag{33}$$

• at very large probe scales $l > l_{\text{IR}}$ gravitational decoherence is absent; a meta-observer testing theory at such scales is dealing with the "full" solution of the Wheeler-de Witt equation, not containing time in analogy with eternal inflation sale (27) in dS spacetime filled with a light scalar field,

• at probe scales $l \lesssim R_H$ purely gravitational decoherence is slow, as it typically takes $t_D \gtrsim R_H$ for the WdW wave function $\psi \sim c_1 \exp(iS_0[a]) + c_2 \exp(-iS_0[a])$ to decohere, if time is measured by the clock associated with the matter degrees of freedom.

Finally, it should be noted that gravity differs from nonrenormalizable field theories described in Sections 2, 3 in several respects, two of which might be of relevance for our analysis: (a) gravity couples to *all* matter degrees of freedom, the fact which might lead to a suppression of the corresponding effective coupling entering in the decoherence factor (30) and (b) it effectively couples to macroscopic configurations of matter fields without any screening effects (this fact is responsible for a rapid decoherence rate calculated in the classic paper [20]). Regarding the point (a), it has been previously argued that the actual scale at which effective field theory for gravity breaks down and gravity becomes strongly coupled is suppressed by the effective number of matter fields N (see for example [36], where the strong coupling scale is estimated to be of the order $M_P/\sqrt{N}$, rather than the Planck mass M_P). It is in fact rather straightforward to extend the arguments presented above to the case of N scalar fields with Z_2 symmetry. One immediately finds that the time scale of decoherence between expanding and contracting branches of the WdW solution is given by

$$H_0 t_d \gtrsim \log \left(\frac{M_P^3}{m N^{1/2} H_0^2} \right)$$

(to be compared with the Eq. (32)), while the single branch decoherence proceeds at time scales of the order

$$H_0 t_d \gtrsim \log \left(\frac{M_P^3}{m^3 N^{3/2}(a_1 - a_2)} \right).$$

For the decoherence between expanding and contracting WdW branches discussed in this Section and for the emergence of cosmological arrow of time, it is important that most of the matter fields are in the corresponding vacuum states (with the exception of light scalars, they are not redshifted away), and the effective N remains rather low, so our estimations remained affected only extremely weakly by N dependence. As for the point (b), macroscopic configurations of matter (again, with the exception of light scalars with $m \ll H_0$) do not yet exist at time scales of interest.

6 discussion

We have concluded the previous Section with an observation that quantum gravitational decoherence responsible for the emergence of the arrow of time is in fact rather ineffective. If the typical curvature scale of the space-time is $\sim R$, it takes at least

$$N \sim \log\left(\frac{M_P^2}{R}\right) \gg 1 \tag{34}$$

efoldings for the quasi-classical WdW wave-function $\psi \sim c_1 \exp(iS_0[a]) + c_2 \exp(-iS_0[a])$ describing a superposition of expanding and contracting regions to decohere into separate WKB branches. Whichever matter degrees of freedom we are dealing with, we expect the estimate (34) to hold and remain robust.

Once the decoherence happened, the direction of the arrow of time is given by the vector $\partial_t = \partial_a S_0 \partial_a$; at smaller spatio-temporal scales than (34) the decoherence factor remains small, and the state of the system represents a quantum foam, the amplitudes $c_{1,2}$ determining probabilities to pick an expanding/contracting WKB branch, correspondingly. Interestingly, the same picture is expected to be reproduced once the probe scale of an observer becomes larger than characteristic curvature scale R. As we explained above, the ineffectiveness of gravitational decoherence is directly related to the fact that gravity is a non-renormalizable theory, which is nearly completely decoupled from the quantum dynamics of the matter degrees of freedom.

If so, a natural question emerges why do we then experience reality as a quasi-classical one with the arrow of time strictly directed from the past to the future and quantum mechanical matter degrees of freedom decohered at macroscopic scales? Given one has an answer to the first part of the question, and the quantum gravitational degrees of freedom are considered as quasi-classical albeit perhaps stochastic ones, its second part is very easy to answer. Quasi-classical stochastic gravitational background radiation leads to a decoherence of

matter degrees of freedom at time scale of the order

$$t_D \sim \left(\frac{M_P}{E_1 - E_2}\right)^2,$$

where $E_{1,2}$ two rest energies of two quantum states of the considered configuration of matter (see for example [37, 38]). This decoherence process happens extremely quickly for macroscopic configurations of total mass much larger than the Planck mass $M_P \sim 10^{-8}$ kg. Thus, the problem, as was mentioned earlier, is with the first part of the question.

As there seems to be no physical mechanism in quantized general relativity leading to quantum gravitational decoherence at spatio-temporal scales smaller than (34), an alternative idea would be to put the burden of fixing the arrow of time on the observer. In particular, it is tempting to use the idea of [39, 40], where it was argued that quasi-classical past → future trajectories are associated with the increase of quantum mutual information between the observer and the observed system and the corresponding increase of the mutual entanglement entropy. Vice versa, it should be expected that quasi-classical trajectories future → past are associated with the decrease of the quantum mutual information. Indeed, consider an observer A, an observed system B and a reservoir R such that the state of the combined system ABR is pure, i.e., R is a purification space of the system AB. It was shown in [39] that

$$\Delta S(A) + \Delta S(B) - \Delta S(R) - \Delta S(A : B) = 0, \tag{35}$$

where $\Delta S(A) = S(\rho_A, t) - S(\rho_A, 0)$ is the difference of the von Neumann entropies of the observer subsystem described by the density matrix ρ_A, estimated at times t and 0, while $\Delta S(A : B)$ is the quantum mutual information difference, trivially related to the difference in quantum mutual entropy for subsystems A and B. It immediately follows from (35) that an apparent decrease of the von Neumann entropy $\Delta S(B) < 0$ is associated with the decrease in the quantum mutual information $\Delta S(A : B)$ < 0, very roughly, erasure of the quantum correlations between A and B (encoded the memory of the observer A during observing the evolution of the system B).

As the direction of the arrow of time is associated with the increase of von Neumann entropy, the observer A is simply unable to recall behavior of the subsystem A associated with the decrease of its von Newmann entropy in time. In other words, if the physical processes representing "probing the future" are possible to physically happen, and our observer is capable to detect them, she will not be able to store the memory about such processes. Once the quantum trajectory returns to the starting point ("present"), any memory about observer's excursion to the future is erased.

It thus becomes clear discussion of the emergence of time (and physics of decoherence in general) demands somewhat stronger involvement of an observer than usually accepted in literature. In particular, one has to prescribe to the observer not only the infrared and ultraviolet "cutoff" scales defining which modes of the probed fields should be regarded as environmental degrees of freedom to be traced out in the density matrix, but also a quantum memory capacity. In particular, if the observer does not possess any quantum memory capacity at all, the accumulation of the mutual information between the observer and the observed physical system is impossible, and the theorem of [39, 40] does not apply: in a sense, the "brainless" observer does not experience time and/or decoherence of any degrees of freedom (as was earlier suggested in [41]).

It should be emphasized that the argument of [39] applies only to quantum mutual information; such processes are possible that the classical mutual information $S_{cl}(A : B)$ increases, whereas the quantum mutual information $S(A : B)$ decreases: recall that the quantum mutual information $S(A : B)$ is the upper bound of $S_{cl}(A : B)$. Thus, the logic of the expression (35) applies to observers with "quantum memory" with exponential capacity in the number of qubits[9] rather than with classical memory with polynomial capacity such as the ones described by Hopfield networks.

References

[1] W. H. Zurek, Decoherence, einselection, and the quantum origins of the classical, Rev. Mod. Phys. 75, 715–775 (2003).

[2] E. Joos, et al. Decoherence and the Appearance of a Classical World in Quantum Theory (Springer Science & Business Media, 2003).

[3] M. Brune, et al. Observing the Progressive Decoherence of the "Meter" in a Quantum Measurement, Phys. Rev. Lett. 77, 4887–4890 (1996).

[4] M. R. Andrews, Observation of Interference Between Two Bose Condensates, Science 275, 637–641 (1997).

[5] M. Arndt, et al. Wave-particle duality of C60 molecules, Nature 401, 680–682 (1999).

[6] J. Friedman, V. Patel, W. Chen, S. Tolpygo, and J. Lukens, Quantum superposition of distinct macroscopic states, Nature 406, 43–6 (2000).

[7] C. H. van der Wal, Quantum Superposition of Macroscopic Persistent-Current States, Science 290, 773–777 (2000).

[8] D. Kielpinski, A Decoherence-Free Quantum Memory Using Trapped Ions, Science 291, 1013–1015 (2001).

[9] D. Vion, et al. Manipulating the quantum state of an electrical circuit, Science 296, 886–9 (2002).

[10] I. Chiorescu, Y. Nakamura, C. J. P. M. Harmans, and J. E. Mooij, Coherent quantum dynamics of a superconducting flux qubit, Science 299, 1869–71 (2003).

[11] L. Hackermüller, et al. Wave Nature of Biomolecules and Fluorofullerenes, Phys. Rev. Lett. 91, 090408 (2003).

[12] L. Hackermüller, K. Hornberger, B. Brezger, A. Zeilinger, and M. Arndt, Decoherence of matter waves by thermal emission of radiation, Nature 427, 711–4 (2004).

[13] J. Martinis, et al. Decoherence in Josephson Qubits from Dielectric Loss, Phys. Rev. Lett. 95, 210503 (2005).

[14] J. R. Petta, et al. Coherent manipulation of coupled electron spins in semiconductor quantum dots, Science 309, 2180–4 (2005).

[15] S. Deléglise et al. Reconstruction of non-classical cavity field states with snapshots of their decoherence, Nature 455, 510–4 (2008).

[16] H.-D. Zeh, The Physical Basis of the Direction of Time (Springer Berlin Heidelberg, Berlin, Heidelberg, 1989).

[17] C. Kiefer, Decoherence in quantum electrodynamics and quantum gravity, Phys. Rev. D 46, 1658–1670 (1992).

[18] C. Anastopoulos, and B. L. Hu, A Master Equation for Gravitational Decoherence: Probing the Textures of Spacetime 24 (2013). 1305.5231.

[19] B. L. Hu, Gravitational Decoherence, Alternative Quantum Theories and Semiclassical Gravity 18 (2014). 1402.6584.

[20] E. Joos, Why do we observe a classical spacetime?, Phys. Lett. A 116, 6–8 (1986).

[21] E. Calzetta and B. L. Hu, Correlations, Decoherence, Dissipation, and Noise in Quantum Field Theory 37 (1995). 9501040.

[22] F. Lombardo and F. Mazzitelli, Coarse graining and decoherence in quantum field theory, Phys. Rev. D 53, 2001–2011 (1996).

[23] J. F. Koksma, T. Prokopec, and M. G. Schmidt, Decoherence and dynamical entropy generation in quantum field theory, Phys. Lett. B 707, 315–318 (2012).

[24] M. Aizenman and R. Graham, On the renormalized coupling constant and the susceptibility in phi 4_4 field theory and the Ising model in four dimensions, Nucl. Phys. B 225, 261–288 (1983).

[25] M. Aizenman, Proof of the Triviality of phi_d4 Field Theory and Some Mean-Field Features of Ising Models for d>4, Phys. Rev. Lett. 47, 1–4 (1981).

[26] M. Aizenman, Geometric analysis of phi 4 fields and Ising models. Parts I and II, Commun. Math. Phys. 86, 1–48 (1982).

---

[9] Number of possible stored patterns is $O(2^n)$, where n is the number of qubits in the memory device.

[27] A. M. Polyakov, Gauge Fields and Strings (CRC Press, 1987).

[28] J. Cardy, Scaling and Renormalization in Statistical Physics (Cambridge University Press, 1996).

[29] A. A. Starobinsky, Stochastic de sitter (inflationary) stage in the early universe, in "Field Theory, Quantum Gravity and Strings", Lecture Notes in Physics, vol. 246 of Lecture Notes in Physics (Springer Berlin Heidelberg, Berlin, Heidelberg, 1986).

[30] A. Starobinsky and J. Yokoyama, Equilibrium state of a self-interacting scalar field in the de Sitter background, Phys. Rev. D50, 6357–6368 (1994).

[31] N. Arkani-Hamed, S. Dubovsky, A. Nicolis, E. Trincherini, and G. Villadoro, A measure of de Sitter entropy and eternal inflation, J. High Energy Phys. 2007, 055–055 (2007). 0704.1814.

[32] R. Woodard, A Leading Log Approximation for Inflationary Quantum Field Theory, Nucl. Phys. B - Proc. Suppl. 148, 108–119 (2005). 0502556.

[33] K. Enqvist, S. Nurmi, D. Podolsky, and G. I. Rigopoulos, On the divergences of inflationary superhorizon perturbations, J. Cosmol. Astropart. Phys. 2008, 025 (2008). 0802.0395.

[34] A. Barvinsky, A. Kamenshchik, C. Kiefer, and I. Mishakov, Decoherence in quantum cosmology at the onset of inflation, Nucl. Phys. B551, 374–396 (1999). 9812043.

[35] C. Kiefer, I. Lohmar, D. Polarski, and A. A. Starobinsky, Pointer states for primordial fluctuations in inflationary cosmology, Class. Quantum Gravity 24, 1699–1718 (2007). 0610700.

[36] G. Dvali, Black Holes and Large N Species Solution to the Hierarchy Problem, Fortsch. Phys. 58, 528–536 (2010). 0706.2050.

[37] M. P. Blencowe, Effective field theory approach to gravitationally induced decoherence, Phys. Rev. Lett. 111, 021302 (2013). 1211.4751.

[38] I. Pikovski, M. Zych, F. Costa, and Č. Brukner, Universal decoherence due to gravitational time dilation, Nat. Phys. 4 (2015). 1311.1095.

[39] L. Maccone, Quantum Solution to the Arrow-of-Time Dilemma, Phys. Rev. Lett. 103, 080401 (2009).

[40] S. Lloyd, Use of mutual information to decrease entropy: Implications for the second law of thermodynamics, Phys. Rev. A39, 5378–5386 (1989).

[41] R. Lanza, A new theory of the Universe, Am. Scholar 76, 18 (2007).

[42] D. Podolsky, On triviality of lambda phi4 quantum field theory in four dimensions. ArXiv:1003.3670.

[43] A. Caldeira and A. Legett, Path integral approach to quantum Brownian motion. Physica A: Stat. Mech. Appl. 121, 587–616 (1983).

APPENDIX 3: OBSERVERS DEFINE THE STRUCTURE OF THE UNIVERSE

Reconciling Quantum Mechanics and General Relativity

Non-technical summary of paper:
The incompatibility between general relativity and quantum mechanics has puzzled generations of scientists starting with Albert Einstein. This paper explains how observers are the key to reconciling these two pillars of modern physics, as well as how they dramatically restructure space itself. It represents a rare case in theoretical physics when the presence of observers drastically changes the behavior of observable quantities themselves not only at microscopic scales but also at very large spatio-temporal scales. And importantly, the paper also provides a possible explanation why the observed dimensionality of spacetime which we live in is four (D = 3 + 1).

Quantum mechanics works exquisitely well in describing nature at the scale of molecules and subatomic particles, while general relativity is peerless in revealing cosmic behavior on the huge scales between the stars. These two theories find numerous practical applications in our everyday life—such as GPS in the case of relativity, and transistors and microprocessors in the case of quantum mechanics. Yet, after almost a century, we lack an understanding how the two are compatible. At the core of this incompatibility is the issue of "non-renormalizability" of quantum gravity (the field that tries to combine the two). As it turns out, the problem of the incompatibility between quantum mechanics and general relativity can be resolved by taking into account something that modern physics has largely ignored till now: the properties of the observers who probe reality.

In physics, it's usually assumed that we're always able to measure the physical state of an object without perturbing it in any way. But in the realm of quantum gravity, this isn't possible. When observers measure the state of space-time foam, the outcomes of their measurements significantly changes when they exchange information—the presence of observers themselves significantly perturbs it. Using simplified language, it matters enormously to the laws of reality that we're here studying and probing it and sharing the results with each other.

This paper has a number of fascinating consequences. First of all, the presence of observers not merely influences but defines physical reality itself. If the reality described by the combination of Einstein's theory of general relativity exists and makes nature operate smoothly, then it also must contain observers in one form or another. Without a network of observers measuring the properties of space-time, the combination of general relativity and quantum mechanics stops working altogether. So it's actually inherent to the structure of reality that observers living in a quantum gravitational universe share information about the results of their measurements and create a cognitive model of it. For, once you measure something, the wave of probability to measure the same value of the already probed physical quantity becomes "localized" or simply "collapses."

This means that if you keep measuring the same quantity over and over again, keeping in mind the result of the very first measurement, you'll see a similar outcome. Similarly, if you learn from somebody about the outcomes of their measurements of a physical quantity, your measurements and those of other observers influence each other—freezing the reality according to that consensus. In this sense, consensus of different opinions regarding the structure of reality defines its very form, shaping the underlying quantum foam.

You might wonder what would happen if there was only one observer in the Universe. The answer depends on whether the observer is conscious, whether he or she has memory about the results of probing the structure of objective reality, whether she builds a cognitive model of this reality. In other words, a single conscious observer can completely define this structure, leading to a collapse of the waves of probability, largely localized in the vicinity of the cognitive model which the observer builds in her mind throughout her life span. As experimental results confirm this, we will be reshaping reality in a way that is long overdue—seeing how intimately we are connected with the structure of the universe on every level.

Journal of **C**osmology and **A**stroparticle **P**hysics
An IOP and SISSA journal

Parisi-Sourlas-like dimensional reduction of quantum gravity in the presence of observers

Dmitriy Podolskiy,[a] **Andrei O. Barvinsky**[b] **and Robert Lanza**[c]

[a]Harvard University,
77 Avenue Louis Pasteur, Boston, MA 02115, U.S.A.
[b]Lebedev Physics Institute, Theory Department
Leninsky Prospect 53, Moscow 117924, Russia
[c]Wake Forest University,
1834 Wake Forest Rd., Winston-Salem, NC 27106, U.S.A.

E-mail: Dmitriy_Podolskiy@hms.harvard.edu, barvin@td.lpi.ru,
robertlanza@outlook.com

Received November 20, 2020
Revised February 4, 2021
Accepted April 28, 2021
Published May 18, 2021

Abstract. We show that in the presence of disorder induced by random networks of observers measuring covariant quantities (such as scalar curvature) $(3+1)$-dimensional quantum gravity exhibits an effective dimensional reduction at large spatio-temporal scales, which is analogous to the Parisi-Sourlas phenomenon observed for quantum field theories in random external fields. After averaging over disorder associated with observer networks, statistical properties of the latter determine both the value of gravitational constant and the effective cosmological constant in the model. Focusing on the dynamics of infrared degrees of freedom we find that the upper critical dimension of the effective theory is lifted from $D_{\rm cr} = 1 + 1$ to $D_{\rm cr} = 3 + 1$ dimensions.

Keywords: modified gravity, initial conditions and eternal universe, alternatives to inflation, inflation

ArXiv ePrint: 2004.09708

https://doi.org/10.1088/1475-7516/2021/05/048

Contents

1 Introduction

Difficulty of establishing connection between general relativity and quantum mechanics has puzzled several generations of theoretical physicists starting with Albert Einstein [1]. At the heart of the problem is perturbative non-renormalizability of naively quantized general relativity [2, 3]. The theory becomes extremely sensitive to the choice of renormalization scheme used. This essentially means that the perturbative control over the behavior of the theory is lost.

This problem resurfaces on multiple levels and within any physical problem involving counting or accounting for quantum gravitational degrees of freedom. For example, in perturbative calculation of gravitational entropy associated with black hole horizon the numerical factor in front of the horizon area acquires infinite perturbative corrections (see [4] for the review). Thus, this prefactor becomes strongly dependent on the choice of regularization scheme entangling the information loss paradox in quantum gravity [5] with the problem of its non-renormalizability. In quantum cosmology, where vacuum energy density essentially determines the expansion rate of the spacetime, perturbative corrections to its value are strongly scale dependent [6]. This implies in turn that even the sign of vacuum energy density is hard to determine with certainty. Again, behavior of the theory in the quantum cosmological setup is not controllable in both ultraviolet and infrared limit.

Starting with the Weinberg's idea of asymptotic safety [7], it has been argued previously that canonical quantum gravity may be non-perturbatively renormalizable, with a UV fixed point. Numerical simulations of Regge-Wheeler simplicial quantum gravity (see [8] and references below), simulations employing dynamical triangulations [9–11] as well as functional renormalization group analysis [12] indeed all point towards the validity of this conclusion.[1] However, given that gravity is the weakest force [13], relying on the existence of a fixed point in a deeply UV regime feels unsatisfactory to us. Indeed, changing matter content of the theory would change in that case the location of the fixed point on its phase diagram possibly even removing it altogether for a specifically chosen matter content. Addressing the problem of non-renormalizability entirely within a perturbative domain, even given the extreme complexity of the problem, thus seems to us a more attractive possibility.

It has also became a common lore that a UV finite theory of gravity such as string theory would automatically guarantee avoidance of the problem of non-renormalizability. For example, counting microstates associated with extremal black hole horizon in string theory produces a correct finite answer 1/4 for the numerical prefactor in black hole entropy [14].[2] On the other hand, it remains poorly understood (see for example [15]) if superstring theories provide *the unique* ultraviolet completion of naively quantized general relativity (GR). Might there be other UV completions which lead to the same controllable behavior in the continuum/infrared limit, completions which we are currently not aware of? In the former case, there should naturally exist a line of arguments which leads to emergence of effective string theoretic representation of the ultraviolet physics from an infrared effective GR setup. It would be desirable to demonstrate explicitly how a stringy behavior naturally emerges from this setup in the UV limit. We believe that the present work identifies a possible new research line along which such arguments can be obtained.

Namely, here we would like to argue that including "observers" which continuously measure such covariant quantities as scalar curvature (i.e., essentially probing the strength of gravitational interaction) and then averaging over disorder associated with a random network of these observers leads to an effectively de Sitter like behavior of the underlying theory of quantum gravity. Consequently, deep infrared behavior of the resulting 3 + 1-dimensional theory is effectively reduced to the one of a 1 + 1-dimensional theory. We identify a possible mapping between degrees of freedom in the original, 3 + 1-dimensional theory of quantum gravity and the ones in the effective 1 + 1-dimensional quantum theory obtained by averaging

---

[1]Interestingly, simulations of both 4-dimensional simplicial quantum gravity and dynamically triangulated 4-dimensional quantum spacetime behave differently above and below this UV fixed point. Namely, an IR phase corresponds to a 4-dimensional AdS-like physics. On the other hand, a UV phase is characterized by a quasi-2-dimensional, branching polymer-like behavior in the UV phase. While this observation is often used in the community as a reason to discard results of these numerical simulations — IR behavior of spacetime we live in is manifestly dS-like rather than asymptotically AdS one — we shall argue below that this difference in behavior above and below the fixed point is actually physical.

[2]String theories are UV complete by construction and contain a scalar-tensor general relativity in their low energy effective actions. In this sense, they represent self-consistent UV completions of the naively quantized general relativity and guarantee UV completeness for the estimations of quantum gravitational effects.

over disorder associated with the presence of "observers" and taking the long wavelength limit. The identified mapping is reminiscent of the celebrated Parisi-Sourlas dimensional reduction known to take place in field theories with global and gauge symmetries in the presence of random external fields [16].

Finally, we also argue that the effective action of the emergent $1 + 1$-dimensional theory coincides with the Liouville scalar theory, i.e., essentially, the theory of two-dimensional quantum gravity [17, 18]. This result has a potential to provide the missing link between naively quantized general relativity and string theory. Importantly, it also explains why observed dimensionality of spacetime which we live in is $D = 3 + 1$.

We deem these observations generally interesting also because the described setup, quantum gravity with disorder, represents a rare case in theoretical physics when the presence of observers drastically changes behavior of observable quantities themselves not only at microscopic scales but also in the infrared limit, at very large spatio-temporal scales. Namely, in the absence of observers the background of the $3 + 1$-dimensional quantum gravity remains unspecified. Once observers are introduced, coupled to the observable gravitational degrees of freedom and integrated out, the effective background of theory becomes de-Sitter like. Rather than being a fundamental constant of the theory, the characteristic curvature of effective cosmological constant is determined by the intrinsic properties of "observers" such as the strength of their coupling to gravity and distribution of observation events across the fluctuating spacetime. Physical observers thus play a critically important role for our conclusions implying a necessity of proper description of observer, observation event and interaction between observers and the observed physical system for theoretical controllability of the very physical setups being probed.

The text of the manuscript is organized as follows. Section 2 is devoted to a numerical study of simplicial Regge-Wheeler (Euclidean) quantum gravity in the presence of random Gaussian field coupled to scalar curvature. We argue that the theory exhibits an analogue of Parisi-Sourlas dimensional reduction after averaging over quenched disorder associated with events of gravitational field probing. In section 3 we represent theoretical arguments explaining results of this study and pointing towards their validity in a continuous Lorenzian quantum theory. The section 5 is devoted to the outline of obtained results and a brief discussion of several analogies of phenomena observed here and the ones realized in condensed matter physics. Finally, appendices include details of numerical simulations of several quantum field theories which we used as a pilot study for the subsequent work on quantum gravity. They also contain a more detailed theoretical derivation of the results of section 3.

2   Parisi-Sourlas-like dimensional reduction in Regge-Wheeler simplicial quantum gravity

Following the approach by Regge and Wheeler [19–21, 27, 28], we consider a pure 4-dimensional Euclidean quantum gravity with a cosmological constant.[3]

---

[3]Regge-Wheeler simplicial gravity [21, 28] might very well be a very distant cousin of the naively quantized general relativity. For example, it is not yet entirely clear if (a) the theory preserves local gauge invariance in the number of dimensions $D > 2$ [29], (b) Euclidean setup critical for the theory is sufficient to capture essentially Lorentzian behavior of true Einstein gravity including, in particular, its gravitational instability, and (c) the theory actually contains a massless spin-2 particle in the spectrum of its low energy perturbations [30]. However, at the moment it remains the best setup which we can use attempting numerical studies of quantum general relativity.

We are interested to determine possible changes in behavior of observables of the theory in the presence of an extra ingredient: von Neumann observers randomly distributed across the fluctuating spacetime and measuring the strength of gravitational self-interaction [22, 23] (please see appendix for the discussion). Observational events associated with their activity can be modeled by the term

$$\sqrt{g}\phi(x)R(x) = 2\sum_{h \supset x} \phi_x \delta_h A_h \qquad (2.1)$$

in the Lagrangian density of the discrete simplicial gravity. In the expression (2.1) the left-hand side of the equality represents a continuum version of the theory with the scalar curvature $\sqrt{g}R(x)$ calculated at the point of spacetime x. The right-hand side represents a corresponding discretized version with the sum running over hinges of simplices crossing the point x and serving as building blocks of spacetime. A_h is the area of the hinge, δ_h is the associated deficit angle defined according to $\delta_h = 2\pi - \sum_{\text{blocks meeting at } \theta} \theta$ and θ is the corresponding dihedral angle. The field ϕ representing von Neumann observers is a source of quenched disorder in the theory which we consider Gaussian distributed in our simulations.

As was briefly mentioned in the Introduction, since the Regge-Wheeler theory possesses a UV fixed point in the number of dimensions $D = 2, 3, 4$ [21], the problem of comparing observables in the presence of disorder (2.1) and without it is reduced to the problem of comparing critical exponents of the theory at the fixed point $k = k_c$. In particular, we were interested in the dependence of the universal critical exponent ν on the background space dimensionality. As usual, we define the critical exponent ν through the average space curvature

$$\frac{\langle \int d^D x \sqrt{g} R \rangle}{\langle \int d^D x \sqrt{g} \rangle} \sim (k_c - k)^{D\nu - 1}, \qquad (2.2)$$

where $k = 1/8\pi G$ and k_c represents the critical point of the theory. Here, the integration is performed over the D-dimensional spacetime in a particular simplicial realization of the lattice theory. The average is taken over the statistical ensemble of simplices, i.e., all realizations in this ensemble. In addition, when observers are introduced, averaging also includes the one over disorder associated with observers (that is, averaging over the ensemble of statistical realizations of observers). In this sense, the denominator of the eq. (2.2) represents the D-volume of the spacetime averaged over realizations of both ensembles, while the numerator — the average scalar curvature. Both integrals are well defined on a finite lattice.

The exponent ν is directly related to the derivative of the beta function for the gravitational constant near the ultraviolet fixed point according to $\beta'(k_c) = -\nu^{-1}$. Namely, in $D = 2 + \epsilon$ space dimensions one has (assuming free gravity with a cosmological constant) [24–26]

$$\frac{1}{8\pi k_c} = \frac{3}{50}\epsilon - \frac{9}{250}\epsilon^2 + \dots, \qquad (2.3)$$

$$\nu^{-1} = -\beta'(k_c) = \epsilon + \frac{3}{5}\epsilon^2 + \dots. \qquad (2.4)$$

To approach the problem in question, we have performed Monte-Carlo simulations of simplicial Euclidean quantum gravity in $D = 4$ space dimensions on hypercubic lattices of sizes $L = 4$ (256 sites, 3840 edges, 6144 simplices), 8 (4096 sites, 6144 edges, 98304 simplices) and 16 (65536 sites, 983040 edges, 1572864 simplices). In all simulations, the topology was fixed to be the one of 4-torus. No fluctuations of topology were allowed. The bare cosmological constant was also fixed to 1 (since the gravitational coupling is setting the overall length scale in the physical problem). To establish efficient thermalization of the system in our numerical experiment (in the absence of disorder) we have investigated behavior of the system at 20 different values of k. For $L = 16$ hyper-lattice 33000 consequent configurations were generated for every single realization of disorder, for $L = 8$ hyper-lattice — 100000 configurations and for $L = 4$ hyper-lattice — 500000 consequent configurations. Obtained dependence of the average curvature (2.2) was then fit to the singular dependence on k to determine the values of critical gravitational coupling k_c and the critical exponent ν. In the absence of disorder (setting all couplings to the disorder field ϕ_k to 0) we found for the $L = 4$ hyper-lattice $k_c = 0.067(3)$, $\nu = 0.34(5)$, for the $L = 8$ hyper-lattice — $k_c = 0.062(5)$, $\nu = 0.33(6)$ and for the $L = 16$ hyper-lattice — $k_c = 0.061(7)$, $\nu = 0.32(9)$; a relatively weak dependence of the fixed point scale k_c on L pointed out towards efficient thermalization of the employed Euclidean lattice system.

We repeated the same procedure for 10000 different realizations of the random disorder ϕ_k. Fitting dependence of the average curvature on k for configurations averaged over disorder, we found the value of k_c (post disorder averaging) to be $k_c \approx 0.03 \pm 0.12$, in principle consistent with $k_c = 0$ (compare with $k_c \approx 0.07$ in the case without disorder). We have found that the value $\nu^{-1} = 0.01 \pm 0.06$ for the $L = 4$ hyper-lattice, $\nu^{-1} = 0.02 \pm 0.05$ for the $L = 8$ hyper-lattice and $\nu^{-1} = 0.02 \pm 0.04$ for the $L = 16$ hyper-lattice (compare with $\nu^{-1} \approx 3$, which holds approximately in the case without disorder).

In principle, both of these observations (vanishing of k_c and ν^{-1}) — but especially the second one — are consistent with a Parisi-Sourlas-like dimensional reduction in the presence of disorder (2.1). Indeed, it has been argued previously (see for example [30]) that $\nu \approx \frac{1}{D-1}$ for large D, while $\nu = \infty$ exactly for $D = 2$. If an analogue of Parisi-Sourlas dimensional reduction holds also for quantum gravity, this naturally implies that the upper critical dimension of gravity ($D = 2$ in the absence of disorder) is lifted to 4 in the presence of a random network of von Neumann detectors performing measurements of scalar curvature.[4] An effective low dimensionality emerging in simulations of simplicial quantum gravity has been previously also reported in [28, 31] where it has been argued that the UV phase of the theory features an effective dimensional reduction with polymer-like behavior of the correlation functions of observables, while its IR physics is smooth with effectively Euclidean AdS (EAdS) background. Vanishing of the critical value k_c after averaging over quenched disorder (2.1) would in turn force one to think that the UV phase becomes the only accessible one across all scales k, naively implying unphysical behavior of the theory in the presence of quenched disorder. We shall argue in the next section that the observed behavior is fully physical and, in sense, a natural one which should be expected from the quantum theory of gravity in the presence of quenched disorder.

Finally, we note in passing that k_c vanishing after averaging over disorder in gravity also seems analogous to a phenomenon which has already been observed in field theories with quenched disorder: for example, the 2nd order phase transition of Ising model (reduced to $\lambda \phi^4$ scalar field theory in the continuum limit) is reached at finite temperature T_c in the absence of disorder and at $T = T_c$ in the presence of random external field [63].

---

[4]One can ask what happens in simplicial Euclidean quantum gravity (with a quenched disorder) at $D_{cr} > 4$ and at $D = 3$? If the analogy with behavior of field theories in external fields holds for gravity completely, we expect the theory of simplicial ($D = 5$)-dimensional quantum gravity with a quenched disorder to be equivalent to a 3-dimensional theory without such disorder etc. On the other hand, Parisi-Sourlas correspondence would break down at $D = 3$ in a similar fashion as it happens in $D = 3$ random field Ising model, see discussion in the appendix. We leave this question to the future study.

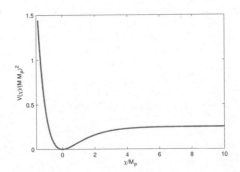

Figure 1. Effective potential $V(\chi) = \frac{M^2 M_P^2}{4}(1 - \exp(-\sqrt{2/3}\chi/M_P))^2$ of the scalar mode $\chi = \sqrt{3/2}M_P \log\left(1 + \frac{R}{M^2}\right)$, where $M = \Delta^{-1/2}M_P$, related to spacetime curvature in the Jordan frame. After averaging over quenched disorder, all not-trivial correlations of gravitational degrees of freedom are represented by correlation functions of χ.

3   Physical origin of possible Parisi-Sourlas-like dimensional reduction in Lorenzian quantum gravity

The observed effect of dimensional reduction in Regge-Wheeler simplicial quantum gravity can be understood (and possibly explained) using the following theoretical arguments.

The continuum limit of the theory (2.1) (assuming that it exists) is expected to correspond to a scalar-tensor Euclidean gravity, where the "dilaton" field ϕ is sufficiently massive, so that its arbitrary configuration in the world volume of the theory can be considered a quenched disorder. The physical idea here is that observers/detectors have a "lifespan" much shorter than the deep infrared scales which we are interested in. Thus, they can be considered as point-like observation "events". The distribution of these "events" is given by the statistical properties of the field ϕ fluctuating across the spacetime. Averaging over the associated quenched disorder (i.e., over the distribution of ϕ), we essentially assume that observers exchange information between each other, and these information exchange processes again happen at time scales much shorter than the infrared scales of interest. Given that, for any given observer the measurement of the gravitational field strength is Bayes-conditioned on the results of similar measurements by other observers in the network. In this sense, the states realized in the effective theory obtained by averaging over distribution of ϕ are "consensus" states.

If the disorder associated with observers is Gaussian, the partition function of the continuum version of the theory is then given by

$$Z = \int \mathcal{D}\phi \int \frac{\mathcal{D}g}{\mathcal{D}f} \exp\left(-\int d^4x\sqrt{g}\left((M_P^2 + \phi)R + \frac{\phi^2}{2\Delta}\right)\right), \tag{3.1}$$

where the integration measure in the path integral over space metric g is assumed to be invariant with respect to arbitrary diffeomorphisms (hence the division by the volume of Diff group $\mathcal{D}f$) and the parameter Δ determines the width of the distribution of the field ϕ describing observers. Note that the field ϕ has the dimension of mass squared, so that the parameter Δ is dimensionless.

Integrating over all possible realizations of ϕ and performing analytic continuation to spacetimes with Lorentz signature,[5] one obtains an effective $f(R)$-theory of gravity with partition function $Z \sim \int \frac{Dg}{Df} \exp(-i \int d^4x \sqrt{-g} f(R))$ and

$$f(R) \sim M_P^2 R + \frac{\Delta R^2}{2}. \tag{3.2}$$

(As an interesting digression, we note that one does not really need to introduce the "bare" gravitational interaction coupling M_P at all if ϕ has a non-zero average over configurations of ϕ; in the latter case, the effective Planck constant simply reads $M_{P,\text{eff}}^2 = \langle \phi \rangle$.)

This theory manifestly admits de Sitter-like solutions for all possible values of its parameters [32]. Such solutions represent dynamical attractors in the phase space of the theory. Indeed, switching from the Jordan frame to the Einstein frame in the $f(R)$-theory of gravity, one finds that the theory (3.2) is effectively equivalent to a theory of gravity coupled to a scalar field

$$\chi \sim \sqrt{3/2} M_P \log \left(1 + \frac{\Delta R}{M_P^2} \right), \tag{3.3}$$

where R is the scalar curvature of spacetime in the original $f(R)$-theory. As always in analysis of an inflationary theory, we are interested in the case of super-Planckian χ, meaning that $\frac{M_P^2}{\Delta} \ll R \ll M_P^2$ (i.e., the dimensionless parameter Δ is very large, so that the distribution of couplings of gravitational detectors to R has a single narrow peak).

The potential of this effective scalar field in the Einstein frame is given by [33–37]

$$V(\chi) \sim \frac{3M_P^4}{4\Delta} (1 - \exp(-\sqrt{2/3}\chi/M_P))^2, \tag{3.4}$$

which reduces to the potential of chaotic inflation at small $\chi \ll M_P$ and a potential quickly approaching a constant asymptotics at $\chi \gg M_P$. In the regime of interest $\Delta \gg 1$ the energy scale is well below M_P.

We also note that the case $\chi < 0$ corresponds to an anti de Sitter-like regime with $\langle R \rangle < 0$ in the Jordan frame, while the regime $\chi > 0$ — to a de Sitter-like physics. For $\chi < 0$, $|\chi| \gg M_P$ corresponds to a collapsing solution in the Einstein frame but anti-de Sitter physics in the Jordan frame with R bounded from below by M_P^2/Δ. Again, physics of the system interestingly depends on the statistical properties of the distribution of observers and observation events in the spacetime. Both the effective value of the Planck constant ($M_P^2 \sim \langle \phi \rangle$) and the one of the effective cosmological constant are determined by statistical properties of the observer networks in this model.

Focusing attention to the case in question with $\chi > 0$, we would like to integrate out sub-horizon fluctuations of the effective field χ (such fluctuations can be considered Gaussian in the first approximation due to applicability of EFT approximation for gravitational degrees of freedom in the UV). Once sub-horizon physics is integrated out, one arrives to the physical picture of an inflationary self-reproducing universe. The only survived "coordinates" here are the number of inflationary efoldings (log of scale factor) and an effective scalar field χ (essentially, a log of scalar curvature in the Einstein frame). In this sense, the originally $(3 + 1)$-dimensional theory becomes effectively 2-dimensional in the infrared. Let us show in details that this is indeed the case using stochastic inflationary formalism [38–49] and

---

[5]The question how such analytic continuation should be performed technically is far from trivial. Here for the sake of simplicity we shall follow the naive prescription for the Wick rotation $t \to -it$.

ignoring gravitational vector and tensor modes which do not contribute to quasi-de Sitter gravitational entropy [50] and thus do not influence strongly infrared dynamics of the theory.

Namely, separating the field χ into the subhorizon and superhorizon parts, one can write:

$$\chi(t,x) = \chi_{IR}(t,x) + \frac{1}{(2\pi)^{3/2}} \int d^3k \cdot \theta(k - \epsilon a H)\left(a_k \phi_k(t) \exp(-ikx) + \text{h.c.}\right) + \delta\chi. \quad (3.5)$$

Here $a(t)$ is the scale factor of de Sitter spacetime, H is the corresponding Hubble constant, $\theta(\ldots)$ is the Heaviside step-function, the modes $\phi_k(t) = \frac{H}{\sqrt{2k}}(\tau - \frac{i}{k})\exp(ik\tau)$ correspond to the Bunch-Davies de Sitter invariant vacuum of a free massless scalar field, $\tau = \int \frac{dt}{a(t)}$ (in what follows, we will be working in the uniform N-gauge), ϵ is a small number such that $\epsilon \ll 1$ (which determines a notation for separating superhorizon modes from the subhorizon ones) and $\delta\chi$ can be neglected in the leading order with respect to H/M_P i.e., when slow roll parameters remain small. This correction encodes the difference between the univorm N gauge and the spatially flat gauge used to derive the correlation properties of the noise $f(\tau,x)$ below [51]. Substituting this decomposition back into the equation of motion for the field χ on de quasi-Sitter background, one obtains the equation for the infrared part of the field χ_{IR}:

$$\frac{d\chi_{IR}}{d\tau} = -\frac{1}{3H^2}\frac{dV}{d\chi_{IR}} + \frac{f(\tau,x)}{H}, \quad (3.6)$$

where a composite operator

$$f(\tau,x) = \frac{\epsilon a H^2}{(2\pi)^{3/2}} \int d^3k \cdot \delta(k - \epsilon a H) \cdot \frac{(-i)H}{\sqrt{2}k^{3/2}}\left[a_k \exp(-ikx) - a_k^\dagger \exp(ikx)\right]$$

has the correlation properties

$$\langle f(\tau,x)f(\tau',x)\rangle = \frac{H^4}{4\pi^2}\delta(\tau - \tau'),$$

if the average is taken over the Bunch-Davies vacuum state. Another very important property of this operator is that its self-commutator vanishes, and thus the equation (3.6) can be considered a stochastic differential equation for the quasi-classical but stochastically distributed long-wavelength field χ_{IR} (from now on, we shall drop the index IR always implying that the infrared, superhorizon part of the field χ is considered).

One then obtains an effective Fokker-Planck equation (see for example appendix D) corresponding to the Langevin equation (3.6) for the probability $P(\tau,\chi)$ to measure a given value of the background/infrared scalar field χ in a given Hubble patch:

$$\frac{\partial P}{\partial \tau} \approx \frac{1}{3\pi M_P^2}\frac{\partial^2}{\partial \chi^2}(VP) + \frac{M_P^2}{8\pi}\frac{\partial}{\partial \chi}\left(\frac{1}{V}\frac{dV}{d\chi}P\right), \quad (3.7)$$

where $\approx$ implies that the equation (3.7) by itself is an approximation (we made a number of simplifications during its derivation such as neglecting the subdominant contribution $\delta\chi$ in the expansion (3.5), assuming slow roll of the field χ and neglecting self-interaction of the field χ at subhorizon scales). We expect that the eq. (3.7) holds on average and only approximately, and model it by including an additional term $F(\tau,\chi)$ to its right-hand side, again quasi-classical but stochastic (see the next section):

$$\frac{\partial P}{\partial \tau} = \frac{1}{3\pi M_P^2}\frac{\partial^2}{\partial \chi^2}(VP) + \frac{M_P^2}{8\pi}\frac{\partial}{\partial \chi}\left(\frac{1}{V}\frac{dV}{d\chi}P\right) + F,$$

Taking into account the smallness of this term, assuming its Gaussianity

$$\langle F(\tau,\chi)F(\tau',\chi')\rangle = A\delta(\tau - \tau')\delta(\chi - \chi')$$

and integrating it out, we finally conclude that the infrared dynamics of the theory (3.1) is being essentially determined by the partition function

$$Z_{IR} = \int \mathcal{D}P \exp(-\mathcal{W}),$$

where the effective action $\mathcal{W}$ of the theory is given by

$$\mathcal{W} = \int d\tau d\chi \frac{1}{A}\left(-\frac{\partial P}{\partial \tau} + \frac{1}{3\pi M_P^2}\frac{\partial^2}{\partial \chi^2}(VP) + \frac{M_P^2}{8\pi}\frac{\partial}{\partial \chi}\left(\frac{dV}{Vd\chi}P\right)\right)^2. \tag{3.8}$$

It is now instructive to use the de Sitter "entropy" S defined according to the prescription $P(\tau,\chi) = \exp(S(\tau,\chi))$ instead of the probability distribution P. One motivation for this substitution is the fact that the distribution function $P(\tau,x)$ converges to

$$P(\tau \to \infty, \chi) \sim \frac{1}{V(\chi)}\exp\left(\frac{3M_P^4}{8V(\chi)}\right)$$

in the limit $\tau \to \infty$, where the expression in the exponent coincides exactly with the gravitational entropy of de Sitter space.[6] As we shall see below, there are other advantages of using S instead of P.

One finds after the substitution

$$\mathcal{W} = \int d\tau d\chi \left(\frac{e^{2S}}{A}\left[-\frac{\partial S}{\partial \tau} + \frac{V}{3\pi M_P^2}(S'' + (S')^2) + (\frac{2V'}{3\pi M_P^2} + \frac{M_P^2}{8\pi}(\log V)')S' + \ldots \right.\right.$$

$$\left.\left.(\frac{V''}{2\pi M_P^2} + \frac{M_P^2(\log V)''}{8\pi})\right]^2 - S\right), \tag{3.9}$$

where prime denotes partial differentiation with respect to the field χ. The appearance of the last term is due to the Jacobian in the measure of functional integration emerging after the change of functional variables.

Substituting the particular form of the potential (3.4) of interest for us to the expression (3.9), we obtain

$$\mathcal{W} = \int d\tau d\chi \left[\frac{e^{2S}}{A}\left(-\frac{\partial S}{\partial \tau} + \frac{M_P}{4\pi}\sqrt{\frac{2}{3}}z\frac{\partial S}{\partial \chi} - \frac{z}{6\pi}\right)^2 + \ldots\right], \tag{3.10}$$

where $z = \exp\left(-\sqrt{\frac{2}{3}}\frac{\chi}{M_P}\right)$. In the quasi-de Sitter limit $S = S_0 + \delta S$, $\delta S \ll S_0$, the potential term in this action coincides with the one of a Liouville-like theory of the "field" S in a two-dimensional spacetime spanned by the coordinates (τ,χ), i.e., the two-dimensional theory of quantum gravity [17, 18] with the "field" S playing the role of the conformal mode of the 2-dimensional spacetime metric.

---

[6]Generally speaking, this holds only for potentials $V(\chi)$ positive for any value of χ. For the potential $V(\chi)$ depicted on the figure 1 this is not the case: as $\chi \to 0$, slow roll approximation breaks down, and inflation ends, which makes the distribution function $P(\tau,\chi)$ non-normalizable, with $P \to 0$ as $\chi \to 0$.

In other words, in the absence of anisotropic stress covariant observables in quantum gravity can be expressed in terms of correlation functions of the scalar curvature (in one-to-one correspondence with the scalar degree of freedom χ in the Einstein frame) according to the prescription

$$\langle \chi^n \rangle \sim \langle (M_P \log \left(\frac{R}{M^2} \right))^n \rangle \sim \int d\chi \cdot \chi^n \int \mathcal{D}S \exp(-\mathcal{W}), \tag{3.11}$$

where the effective action in the path integration (3.11) is given by the expression (3.10). Therefore, the limit $\log a \to \infty$ of the integrand in eq. (3.11) can be thought of as a ground state of the theory of gravity (3.2).

To formalize the map (3.11) a bit clearer, we can write averages of any observable $\langle \mathcal{O} \rangle(t)$ at the time $t \sim H^{-1} \log(a)$ as

$$\langle \mathcal{O}(\chi) \rangle(t) = \int d\chi \, \mathcal{O}(\chi) \, P(\chi, t), \tag{3.12}$$

where the partition function P satisfies the Fokker-Planck equation

$$E(\dot{P}, \partial_\chi P, \chi) \equiv -\dot{P} + \hat{H}P = 0. \tag{3.13}$$

We thus have a chain of transformations

$$\langle \mathcal{O}(\chi) \rangle(t) = \int d\chi \, \mathcal{O}(\chi) \, P(\chi, t) = \int d\chi \, \mathcal{O}(\chi) \int D\bar{P} \, \bar{P}(\chi, t) \, \delta[\, E(\dot{\bar{P}}, \partial_\chi \bar{P}, \chi)\,], \tag{3.14}$$

where the Jacobian for transformation between E and P is disregarded for simplicity. The functional delta-function in the last integral on the right is effectively regulated by the small parameter $A \to 0$ according to

$$\delta[\, E(\dot{\bar{P}}, \nabla \bar{P}, \chi)\,] = \prod_{t,\chi} \delta\Big(E(\dot{\bar{P}}(t,\chi), \partial_\chi \bar{P}(t,\chi), \chi) \Big) \equiv \prod_x \delta\Big(E(\nabla \bar{P}(x), \nabla \bar{P}(x), \chi) \Big)$$

$$= \exp\Big(-\frac{1}{A} \int d^2x \, E^2(\nabla \bar{P}(x), \nabla \bar{P}(x), \chi) \Big). \tag{3.15}$$

Here we introduced 2D coordinates $(x_0, x_1) = (\log(a), \chi)$, and $\nabla = \partial_0, \partial_1$ collects all 2D derivatives. Thus we have (dropping bar over functional integration variable P)

$$\langle \mathcal{O}(\chi) \rangle(t) = \int D\bar{P} \, \exp\Big(-\frac{1}{A} \int d^2x \, E^2(\nabla \bar{P}(x), \nabla \bar{P}(x), \chi) \Big) \int d\chi \, \mathcal{O}(\chi) \, \bar{P}(\chi, t)$$

$$= \int DP \, \exp\Big(-\frac{1}{A} \int d^2x \, E^2(\nabla P(x), \nabla P(x), \chi) \Big) \int d^2y \, \mathcal{O}(y^1) \, P(y) \, \delta(y^0 - t).$$

The probability P to measure a given value of χ in a given Hubble patch is then re-parametrized as $P \sim \exp(S)$, and S become the field variable of interest for us. (Note that in the quasi-de Sitter limit S coincides with the gravitational entropy of de Sitter space in the limit $\log(a) \to 0$.) The mapping dictionary of duality between 2D side and 4D side for the action in quantum measure and for observables of interest is then defined according to the prescription

$$\exp\Big(-\frac{1}{A} \int d^2x \, E^2(\nabla \bar{P}(x), \nabla \bar{P}(x), \chi) \Big) \quad \Leftrightarrow \quad \exp\Big(-\text{Polyakov action} \Big) \tag{3.16}$$

$$\mathcal{O}(\chi) \quad \Leftrightarrow \quad \int d^2y \, \mathcal{O}(y^1) \, P(y) \, \delta(y^0 - t), \tag{3.17}$$

The physical reason why the dimensional reduction has effectively realized in the theory (3.2) and its analytic continuation to Lorentzian spacetimes is simple. Once the dynamics of relevant degrees of freedom is coarse-grained to comoving spatio-temporal scales $\sim H_0^{-1} \sim \Lambda^{-1/2}$,[7] the global structure of spacetime is represented by a set of causally unconnected Hubble patches. Expectation values and correlation functions of the field χ are determined by a stochastic process generated by the Langevin equation (3.6). The values of χ in different Hubble patches are completely independent of each other, and thus the spatial dependence of χ becomes largely irrelevant.

We have argued that the 4-dimensional gravity with a quenched "dilaton" χ becomes reduced to an effectively two-dimensional theory in the deep infrared limit (of large spatio-temporal coarse-graining), where coordinate mapping of the fluctuating spacetime is given in terms of the number of efoldings $\tau = \log a$ and the effective scalar degree of freedom χ related to the large-scale curvature of spacetime in the Einstein frame according to the prescription (3.3). We emphasize that the physical scales at which this description becomes efficient coincide and exceed the scales of eternal inflation from the point of view of a subhorizon observer, thus effectively regularizing the structure of the theory in this deep IR limit. Tensor and vector degrees of freedom present in the metric for the subhorizon observer are effectively integrated away and do not contribute to the infrared structure of the correlation functions of observables in the theory. When the probe scale approaches the cosmological horizon scale, this effectively 2D physics has to be matched to an effective 4D field theory description of gravitational degrees of freedom. It is quite clear from the setup how it has to be done physically. Namely, effective subhorizon 4D degrees of freedom including vector and tensor ones are propagating on the stochastic background with large scale statistical properties effectively determined by the Liouville physics described above.

4   Fokker-Planck equation and its extensions in the two-noise model

In this section, we shall derive the effective action (3.8) used above, albeit in a schematic fashion, and estimate dependence of the parameter A in (3.8) on δ and slow roll parameters.

As was discussed previously, the inflationary Fokker-Planck equation holds its canonical celebrated form (3.7) only in the regime $\delta \to 0$, $\epsilon_H \to 0$, which does not necessarily hold anywhere except very close to the de Sitter spacetime geometry. Moreover, even for geometries globally close to dS_4 one might be interested in behavior of the IR effective theory under different values of parameter δ separating IR and UV physics. At this point, one would only be aware of the fact that the theory approaches the regime $A \to 0$ with Fokker-Planck-like dynamics of $P(\chi, N)$ at $\delta \to 0$, which is entirely independent of δ. In short, we would like to derive extension of this equation which would hold to first order in slow roll parameters ϵ_H, η_H and, ideally, to order in δ higher than first.

First of all, one notes that the one-noise stochastic model for the infrared dynamics of the scalar field in the inflationary spacetime (3.6)–(3.7) cannot be used for this derivation as it produces manifestly non-local results, see appendix E. This non-locality stems from the presence of additional degree of freedom which is integrated out to obtain the effective theory (3.6)–(3.7) in the case of generic ϵ_H, δ. It can be shown that this degree of freedom

---

[7] As we are interested in the continuum limit of the theory (2.1), it is natural to study exactly this case.

can be accounted for if we consider a two-noise model similar to the one introduced in [40]:

$$\frac{d\Phi}{dN} = \frac{v}{H} + \sigma, \tag{4.1}$$

$$\frac{dv}{dN} = -3v - H^{-1}\frac{\partial V}{\partial \Phi} + \tau, \tag{4.2}$$

where

$$\sigma(N, \mathbf{x}) = \frac{1}{H} \int \frac{d^3k}{(2\pi)^{3/2}} \delta(k - \delta aH) \left(a_k \phi_k e^{-i\mathbf{k}\mathbf{x}} + \text{h.c.} \right) \tag{4.3}$$

$$\tau(N, \mathbf{x}) = \frac{1}{H} \int \frac{d^3k}{(2\pi)^{3/2}} \delta(k - \delta aH) \left(a_k \dot{\phi}_k e^{-i\mathbf{k}\mathbf{x}} + \text{h.c.} \right) \tag{4.4}$$

The expressions for the modes ϕ_k, $\dot{\phi}_k$ have to be derived under assumption of finite (but small ϵ_H). Importantly, to the first order in small roll parameters we can keep ϵ_H constant (see appendix E). The equation for the modes $u_k = a\phi_k$ then has the form

$$0 = u_k'' + \left(k^2 - \frac{a''}{a} + m^2 a^2 \right) u_k = u_k'' + \left(k^2 - \left(2 - \epsilon_H - \frac{m^2}{H^2} \right) H^2 a^2 \right), \tag{4.5}$$

where m is the effective mass of the scalar field. To the leading order, we have $\epsilon_H \approx \frac{m^2}{3H^2}$, and thus $2 - \epsilon_H - \frac{m^2}{H^2} \approx 2 - 4\epsilon_H = 2(1 - \epsilon_H)$. One the other hand, again, $a \approx -\frac{1}{H\eta}(1 + \epsilon_H) + \mathcal{O}(\epsilon_H^2)$, and we finally obtain that (to the leading linear order in slow roll parameters ϵ_H) the field u_k satisfies the free massless field equation

$$u_k'' + \left(k^2 - \frac{2}{H^2\eta^2} \right) = 0, \tag{4.6}$$

with its properly normalized solution given by

$$u_k(\eta) = -\frac{1}{\sqrt{2k}} \left(1 - \frac{i}{k\eta} \right) \exp(-ik\eta) + \mathcal{O}(\epsilon_H^2). \tag{4.7}$$

Thus, to the first order in ϵ_H the mode ϕ_k is given by

$$\phi_k(\eta) = \frac{H(1 + \epsilon_H)}{\sqrt{2k}} \left(\eta - \frac{i}{k} \right) \exp(-ik\eta) + \mathcal{O}(\epsilon_H^2) \tag{4.8}$$

Differentiating this expression with respect to world time $t = \int \frac{dN}{H}$ we find

$$\frac{d\phi_k(\eta)}{dt} = \frac{ikH^2\eta^2}{\sqrt{2k}} e^{-ik\eta} - \frac{\epsilon_H H^2}{\sqrt{2k}} \left(\eta - \frac{i}{k} \right) e^{-ik\eta} + \mathcal{O}(\epsilon_H^2). \tag{4.9}$$

The correlation functions of the noise terms $\sigma(N)$ and $\tau(N)$ (as usual, we are interested only in the behavior of correlation functions in the same Hubble patch parametrized by the same "coarse-grained" spatial point $\mathbf{x}$) are in turn found to be

$$\langle \sigma(N)\sigma(N') \rangle \approx \frac{H^2}{4\pi^2} \delta(N - N') \left(1 + 3\epsilon_H + \delta^2 \right), \tag{4.10}$$

$$\langle \tau(N)\tau(N') \rangle \approx \frac{\delta^2 H^4}{4\pi^2} \delta(N - N') \left(\delta^2 + 2\epsilon_H \right) \tag{4.11}$$

$$\langle \tau(N)\sigma(N') \rangle \approx \delta(N - N') \frac{H^3}{4\pi^2} (\epsilon_H + \delta^2 - i\epsilon_H\dot{\delta}) \tag{4.12}$$

(the latter two correlation functions vanish in the limit $\delta \to 0$), and the correlation function $\langle \sigma(N) \tau(N') \rangle$ is related to (4.12) by complex conjugation. Mixed $\tau\sigma$ correlators also end up suppressed by either powers of δ or ϵ_H. Note in this respect that one has to be careful taking one of the limits $\delta \to 0$ or $\epsilon_H \to 0$ first (compare for example to [40]). Two limiting cases are of special interest:

(a) **Quasi-de Sitter limit.** $\epsilon_H \ll \delta^2 < 1$, the case considered in the appendix E with ϵ_H negligible but keeping all orders in δ

$$\langle \sigma(N)\sigma(N') \rangle \approx \frac{H^2}{4\pi^2}(1 + \delta^2)\delta(N - N'),$$

$$\langle \tau(N)\tau(N') \rangle \approx \frac{\delta^4 H^4}{4\pi^2}\delta(N - N')$$

$$\langle \tau(N)\sigma(N') \rangle \approx \frac{\delta^2 H^3(1 + i\delta)}{4\pi^2}\delta(N - N')$$

and

(b) **"Deep IR physics" or Nambu-Sasaki limit.** $\delta^2 \ll \epsilon_H \ll 1$:

$$\langle \sigma(N)\sigma(N') \rangle \approx \frac{H^2}{4\pi^2}(1 + 3\epsilon_H)\delta(N - N'),$$

$$\langle \tau(N)\tau(N') \rangle \approx \frac{2\delta^2 \epsilon_H H^4}{4\pi^2}\delta(N - N') \approx 0,$$

$$\langle \tau(N)\sigma(N') \rangle \approx \frac{\epsilon_H H^3}{4\pi^2}\delta(N - N')$$

While both cases are very illustrative and somewhat similar (specifically, in the regime $\delta \ll 1$), here for our purposes we will focus on the first one, in which functional integrations are simplified greatly. The opposite case (b) is considered in relative depths in [40] and will be discussed in more details in a subsequent work.

To derive the Fokker-Planck equation and corrections to it, we follow the path integral approach outlined in the appendix D. It can be seen easily that the diffusion matrix associated with the correlation properties of the noises is singular in the quasi-de Sitter limit (a), and the functional integration measure for the noise terms has the form

$$Z_{\text{noise}} = \int \mathcal{D}\sigma \mathcal{D}\tau \exp\left(-\frac{1}{2}\int dN\, f^T D^{-1} f\right), \tag{4.13}$$

where

$$D = \begin{pmatrix} \frac{H^2(1+\delta^2)}{4\pi^2} & \frac{H^3\delta^2(1+i\delta)}{4\pi^2} \\ \frac{H^3\delta^2(1-i\delta)}{4\pi^2} & \frac{\delta^4 H^4}{4\pi^2} \end{pmatrix} \tag{4.14}$$

and $f^T = (\sigma, \tau)$. The matrix D is manifestly singular, and the noises σ and δ are correlated. Calculating eigenvectors and eigenvalues of the matrix D, we find that

$$\tau = -H\sigma \frac{i\delta^2}{i + \delta}, \tag{4.15}$$

while the non-trivial contribution into (4.13) is given by the combination $H^{-1}\tau + \frac{i(\delta - i)}{\delta^2}\sigma$. Since the matrix D is singular only in the limit of vanishing slow roll parameters $\epsilon_H \to 0$, it should be kept in mind that its eigenvector corresponding to the zero eigenvalue really introduces a constraint on the dynamics of v and Φ.

The correlation functions of Φ and v can in turn be obtained by integrating over the measure

$$
F = \int \mathcal{D}\sigma \mathcal{D}\tau \mathcal{D}v \mathcal{D}\Phi \mathcal{D}\lambda \mathcal{D}\mu \exp \left(\int dN \left(i\lambda \left(\frac{\partial v}{\partial N} + 3v + \frac{\partial V/\partial \Phi}{H} - \tau \right) + \right. \right.
$$
$$
\left. \left. + i\mu \left(\frac{\partial \Phi}{\partial N} - \frac{v}{H} - \sigma \right) \right) \right) Z_{\text{noise}}.
\tag{4.16}
$$

As all integrations (with the exception of the integration over Φ) are Gaussian, they can be explicitly taken revealing

$$
F = \int \mathcal{D}\Phi \mathcal{D}v \exp(-S),
$$

where

$$
S = \int dN \mathcal{L} = \frac{1}{2} \int dN \frac{2\pi^2}{H^2(1 + \delta^2 + \delta^4)} \left(\frac{1 + i\delta}{\delta^2} \left(\frac{\partial \Phi}{\partial N} - \frac{v}{H} \right) + \right.
$$
$$
\left. + \frac{1}{H} \left(\frac{\partial v}{\partial N} + 3v + \frac{1}{H} \frac{\partial V}{\partial \Phi} \right) \right)^2.
\tag{4.17}
$$

On top of this effective action there is a constraint present in the system (the one corresponding to the vanishing eigenvalue of the matrix (4.14)):

$$
\frac{\delta^2}{1 - i\delta} \left(\frac{\partial \Phi}{\partial N} - \frac{v}{H} \right) = \frac{1}{H} \left(\frac{\partial v}{\partial N} + 3v + \frac{1}{H} \frac{\partial V}{\partial \Phi} \right)
\tag{4.18}
$$

Solving it for $\frac{\partial \Phi}{\partial N} - \frac{v}{H}$ and substituting back into the action (4.17), we obtain the effective theory of the field v:

$$
Z = \int \mathcal{D}v \mathcal{D}\Phi \exp \left(-\frac{2\pi^2 K}{H^4} \left(\frac{\partial v}{\partial N} + 3v + \frac{1}{H} \frac{\partial V}{\partial \Phi} \right)^2 \right),
\tag{4.19}
$$

where $K = (1 + \delta^2)^2 / (\delta^8 \cdot (1 + \delta^2 + \delta^4))$ (in this representation, Φ is considered an external field which we average out).

The conjugate momentum for the field v is given by

$$
p_v = \frac{\pi^2 K}{H^4} \left(\frac{\partial v}{\partial N} + 3v + \frac{1}{H} \frac{\partial V}{\partial \Phi} \right)
\tag{4.20}
$$

and the Hamiltonian of the theory (4.19) is

$$
\mathcal{H}_v = -\frac{H^4}{\pi^2 K} p_v^2 - 3v - \frac{1}{H} \frac{\partial V}{\partial \Phi}.
\tag{4.21}
$$

The Fokker-Planck equation for the probability $P(v, N)$ to measure a given value of v in a given Hubble patch is then obtained by writing down $\frac{\partial P(v,N)}{\partial N} = \mathcal{H}_v(p_v, v) P(v, N)$ and promoting conjugate momentum p_v into a differential operator according to the usual prescription $p_v = -\partial_v$, see appendix D.

An important conclusion is that the theory with the Hamiltonian (4.21) is generally unstable, with a run-away behavior of the probability $P(v, N)$. It can be shown that this conclusion survives in the general case, independent on relations between ϵ_H and δ (as we shall demonstrate in the follow-up work): namely, the run-away behavior is associated with the behavior of the probability distribution P as a function of p_v, while its behavior as a function of p_Φ and Φ remains stable. This is a reflection of general instability of de Sitter space [52]. One and perhaps the only way to deal with this instability is to set a general constraint $p_v = 0$ (i.e., to choose initial conditions for the physical system in this rather special way). Then, from the constraint (4.18) we obtain

$$v = H\frac{\partial \Phi}{\partial N} \tag{4.22}$$

and substituting it back into the effective action of the theory (4.17), we finally obtain the theory with Langrangian

$$\mathcal{L} = \frac{2\pi^2}{H^4(1+\delta^2+\delta^4)}\left(\frac{\partial}{\partial N}\left(H\frac{\partial \Phi}{\partial N}\right) + 3H\frac{\partial \Phi}{\partial N} + \frac{1}{H}\frac{\partial V}{\partial \Phi}\right)^2. \tag{4.23}$$

It is then straightforward to show that the theory (4.23) produces the canonical Starobinsky-Fokker-Planck equation with a singular correction $\sim A\delta(\Phi - \Phi')$ originating from the first term in parentheses in (4.23) and $A \sim H^2$.

5   Conclusion

Numerical and theoretical analysis of non-renormalizable field theories and 4-dimensional quantum gravity performed here shows that introducing a network of observers distributed in the world volume of the theory and continuously measuring scalar curvature leads to a drastic non-perturbative restructuring of the Hilbert space of the underlying theory significantly changing its infrared structure. Perhaps, most importantly, integrating out observers induces a de Sitter-like background of the theory. This in turn completely determines the infrared structure of correlation functions of physical observables. Moreover, introducing random ensembles of observers renormalizes the gravitational constant as well. Namely, while the magnitude of the cosmological constant is determined by the standard deviation of the observer coupling to the scalar curvature across the ensemble, the renormalized Planck mass squared is proportional to the average observer coupling to the scalar curvature, where the corresponding average and standard deviation are calculated using the distribution function of observer couplings. Thus, the induced cosmological constant and the gravitational constant are determined by the properties of observers — the distribution of observation events in the world volume of the theory and coupling strength between observers and gravitational degrees of freedom.

On the one hand, restructuring of the Hilbert space of the gravity coupled to observers is similar to the phenomenon of Anderson-like localization in disordered media. On the other hand, it is characterized by an effective dimensional reduction close in spirit to the celebrated Parisi-Sourlas dimensional reduction observed in several field theories in the presence of random external field (such as continuum limit of RFIM — random field Ising model).

In the case of gravity, this Hilbert space restructuring can be roughly characterized as follows. It is known by now that 4D simplicial Euclidean quantum gravity admits a UV fixed point at a particular value of $G_c = 1/(8\pi k_c)$, with a UV, strongly coupled phase at $k < k_c$ and an IR, weakly coupled phase, which is realized at $k > k_c$. The strongly coupled phase admits a 4-dimensional Euclidean Anti de Sitter like behavior of the ground state of gravity (i.e., expectation value of the scalar curvature remains negative in this phase) and a non-trivial infrared dynamics of correlation functions of observables. On the other hand, the weakly coupled phase (which should be the one of physical interest as the real world gravity is weakly coupled!) seems to feature a quasi-2-dimensional branching polymer-like behavior without any smooth background geometry in the IR. This was previously interpreted as an absence of a proper continuum limit of the theory in this regime. Instead, we believe that this behavior is physical: in the weakly coupled regime the ground state of the theory admits a dS-like physics, and a 2-dimensional branching polymer, self-reproducing behavior of observables is really nothing but a Wick-rotated equivalent of the eternal inflation happening on this dS-like background.

Quenched disorder associated with random networks of observers measuring the strength of gravitational interaction clears up the mist somewhat in this respect: it seems to move the critical point of the theory towards $k_c \to 0$ implying that the only accessible phase of the theory is the one of weakly coupled gravity. Our theoretical analysis further shows that the nature of the effective dimensional reduction $(4D \to 2D)$ in the presence of quenched disorder is associated with the fact that infrared dynamics of observables in the eternally inflating Universe is determined by the probability $P(\chi, \log(a)) = \exp\left(S(\chi, \log(a))\right)$ to measure a given value of the effective "inflaton" χ in a given Hubble patch. In other words, all correlation functions of physical observables are entirely determined by the structure of $P(\chi, \log(a))$. We find this observation rather interesting.

The discussed phenomenon might also explain what should really be understood by the "continuum limit" of non-renormalizable quantum gravity. Indeed, it is generally accepted that the formal continuum limit of non-renormalizable quantum field theories (including in principle 4-dimensional quantum general relativity, which might be non-perturbatively renormalizable) does not exist. Nevertheless, it is still possible to make a number of conclusions regarding the physical properties of such theories in the large-scale/infrared limit. To a degree, the way how to do it can be understood using the correspondence between relativistic quantum field theories and the corresponding statistical classical field theories obtained from the former by Wick rotation [53]. According to this correspondence, the classical statistical counterpart of a renormalizable quantum field theory describes behavior of the order parameter of a classical statistical system in a vicinity of a second order phase transition. At temperatures close to T_c the correlation length $\xi \sim |T - T_c|^{-\alpha}$ of the relevant degrees of freedom approaches infinity, which makes it possible to describe correlation functions of observables in terms of a small number of continuous order parameters only. Similarly, the statistical physics counterpart of a *non-renormalizable* quantum field theory describes a vicinity of *a first order* phase transition, when the correlation length ξ of physical degrees of freedom remains finite at all accessible values of thermodynamic potentials. The process of measuring the physical state of the field in an equivalent QFT can be thought of as an insertion of a projection operator in the world volume of the theory at a point of spacetime, where and when the measurement/observation is performed. In the statistical mechanical counterpart, such insertion is akin to an introduction of a heavy, "quenched" impurity in the spatial volume of the classical thermodynamic system, with elementary excitations of the order parameter(s) scattering against it. It can thus be expected that the network of von Neumann observers in a QFT is reminiscent of an ensemble of impurities introduced into the classical system described by a statistical counterpart of the theory. It is well known that in the vicinity of a first order phase transition, such impurities serve as nucleation centers for bubbles of the true phase [54]. When coupling constants of the von Neumann detectors to the field are sufficiently large, bubble nucleation process proceeds ad infinitum in a quasi-continuum limit

$T \to T_c$. Correspondingly, in a QFT, the vacuum state remains largely inhomogeneous even in the limit of Langevin time $\tau \to \infty$; the resulting state is also strongly dependent on the particular location of inserted operators describing observation events. Thus, the structure of the Hilbert space of a non-renormalizable quantum field theory is largely determined by "localization properties" of the effective potential of impurities inserted into an equivalent statistical mechanical system.

To conclude, two observations presented here point out towards a possible $4D \to 2D$ dimensional reduction of quantum gravity (at least in the infrared) in the presence of random networks of observers: (a) lattice simulations of simplicial Regge-Wheeler Euclidean gravity in the presence of disorder showing that after averaging over such disorder the critical exponent(s) of the theory change as if the effective dimensionality of the theory changes from $D = 1 + 3$ to $D = 1 + 1$, and (b) theoretical analysis of quantum (Lorenzian) general relativity in the presence of quenched disorder. Although these two separate approaches lead to the same conclusion regarding the physical system in question, ultimately only numerical simulations Lorentzian quantum gravity will allow to reconcile the two approaches. We hope to return to this subject in our future studies.

A   Notes on von Neumann observers

Our notation of a physical observer is similar to the one used by Unruh [22] and DeWitt [23]. Namely, in field theoretic setups discussed below we assume that the detector is moving along a world line $x = x(\tau)$, where τ is its proper time, and interacting with the physical field as described by the monopole interaction term in the Lagrangian of the joint system "detector-field" $I(\tau)\phi(x(\tau))$ (note that the monopole coupling strength I of the detector to the field ϕ might itself depend on the proper time of the detector). In the case of gravity, we consider a similar interaction term coupling the monopole detector with the covariant term in the Einstein-Hilbert action, e.g., Ricci scalar, writing this interaction term as $I(\tau)R(x(\tau))$. Further, since we are interested in extremely long spatio-temporal scales (possible infrared limit of the theory), we assume that the lifetime of any particular detector, i.e., the proper total length of its trajectory $L = \int d\tau$ is much shorter than any physical scales of interest. Thus, world lines of monopole detectors can be essentially represented by delta-functions in the world volume of the theory. In the main text, we consider the situation when a large number of such detectors is placed in a world volume of the theory, characterized by varying strengths $I(x)$. The random ensemble of the monopole coupling strengths $I(x)$ of these detectors can be essentially considered as a single "frozen" realization of a scalar field, which we denote in the text as ϕ. Finally, we allow detectors to "consciously" exchange information about the measurements of the Ricci scalar between each other and can thus average over realizations of ϕ assuming ergodicity of their distribution in the spacetime (i.e., assuming that the average over realizations of the field ϕ is equivalent to the average over the 4-volume of the theory); in this sense, we denote these detectors as von Neumann-Wigner observers.

B   Parisi-Sourlas dimensional reduction in non-renormalizable field theories in the presence of observer networks

Quantum gravity can be thought of as a "quantum field theory" with an infinite dimensional gauge symmetry (Lorentz groups of local coordinate transformations in every point of spacetime) [2, 3]). It naturally makes sense to consider simplified models of the same phenomenon which we described above by making dimensionality of the symmetry group finite and recall how Parisi-Sourlas dimensional reduction (due to the presence of random observer networks) emerges in the class of theories with finite-dimensional symmetries.

The first analyzed model of interest is a non-renormalizable scalar field theory with Z_2 global symmetry in $D = 5$ and 6 spacetime dimensions, with the Lagrangian density of the form $\mathcal{L} = \frac{1}{2}(\partial\phi)^2 - \frac{1}{2}m^2\phi^2 - \frac{1}{4}\lambda_0\phi^4 - \ldots$, where $\ldots$ denotes higher-order terms in powers of the scalar field ϕ and its spacetime derivatives ∂. It is well-known that this theory is trivial [48, 55, 56], which implies that all its critical exponents coincide with the ones given by the mean field theory approximation (with logarithmic corrections in 4 dimensions [56]). In other words, if the number of spacetime dimensions $D > 4$, the quantum effective action of the ϕ^4 theory in the continuum limit can be well described by the one of a free massive scalar field theory with an effective mass dependent on the bare coupling λ_0.

Consider a system of von Neumann detectors excited during the interaction events with quanta of the field ϕ [57, 58]. As usual, such detectors with monopole moments $J_i = J_i(t, x)$ can be modeled by terms in the Lagrangian density of the theory linear in the field variable ϕ as

$$Z = \int \mathcal{D}\phi \exp\left(-i\sum_j \int d^D x \left(\frac{1}{2}(\partial\phi)^2 - \frac{1}{2}m_0^2\phi^2 - \frac{1}{4}\lambda_0\phi^4 + J_j\phi\right)\right).$$

Here the sum $\sum_i$ runs over detectors distributed in the world volume of the theory. It is often convenient to think of sources J_i as a second, extra massive scalar field (with a suppressed kinetic term). We are specifically interested in the case when a very large number of such von Neumann detectors randomly located in the world volume of the theory is present. The detectors are characterized by random couplings J_i to the field ϕ.

It can be seen straightforwardly that the physical setup described here is equivalent to the one realized in a quantum $\lambda\phi^4$ theory in a random external field or its discrete version, the random field D-dimensional Ising model (RFIM) well studied in literature, see for example [59–62]). A celebrated result by Parisi and Sourlas [16] states that the infrared behavior of RFIM is equivalent to the one of a similar theory (Ising model) in the absence of random external field but living in $(D-2)$ dimensions. Namely, the most infrared divergent terms present in the perturbative expansion of the generating functionals of the two theories $((D-2)$-dimensional IM and D-dimensional RFIM) coincide term by term. While the Parisi-Sourlas correspondence for Ising model breaks down for $D < 3$ [60], it has been shown to hold universally for $D \geq 4$.

Of especial interest for us is the observation that the presence of a large number of von Neumann detectors drastically changes the structure of the Hilbert space of the theory. For $D = 4$ and 5 the Hilbert space can no longer be approximated by the mean field-theoretic partition function once the disorder associated with the external field is averaged out. This is reflected in the change of critical exponents of the theory as well as correlation functions of the field.

The magnitude of the change in the structure of Hilbert space of the theory can be assessed by numerical simulations. Those have been recently done in [60] for $D = 3$ random field Ising model (RFIM), in [59, 61] for $D = 4$ RFIM and in [62] for $D = 5$ RFIM. We have also performed lattice numerical simulations of RFIM and reproduced the known results for comparison of $D = 4, 5$ RFIM with pure Ising model in $D = 2, 3$ dimensions. Similar to [59–61] we have exploited the fact that RFIM achieves phase transition at zero temperature [63] and as such, it is sufficient to focus on the physics of the ground state of the theory. For numerical simulations, we have used minimum cost-flow algorithm [64, 65].

In addition, we have also performed simulations of $D = 6$ RFIM and compared its behavior with the one of pure Ising model in $D = 4$ dimensions. As was expected, Parisi-Sourlas dimensional reduction was observed in 5-dimensional and 6-dimensional RFIM, with the critical exponents of $D = 4$ RFIM deviating from the ones of 2-dimensional Ising model due to the known breakdown of dimensional reduction mechanism in lower dimensions. Estimating critical exponents for $D = 6$ RFIM we were unable to detect logarithmically weak corrections to the mean field approximation.

To confirm universality of Hilbert space restructuring in quantum field theories due to the presence of networks of observers/observation events, we have also performed numerical simulations of Z_2 gauge theories. Those included theories in $D = 2 + 1$ and $3 + 1$ spacetime dimensions [66–68], $D = 4+1$ and $5+1$ spacetime dimensions (here $+1$ denotes the dimension with periodic boundary conditions). Von Neumann observers were modeled by a scalar degree of freedom coupled to the Z_2 gauge field with the resulting free energy of the theory given by

$$F = \sum_{i,j,k,l} \sigma_{ij}\sigma_{jk}\sigma_{kl}\sigma_{li} + \sum_{i,j,n} g_n \tau_i \sigma_{ij} \tau_j, \tag{B.1}$$

where the couplings g_n (strengths of detectors' couplings to Z_2 gauge field) and locations of insertions of the quenched disorder elements (observation events) were considered random and Gaussian-distributed. Again, we have observed effective dimensional reduction in the random field Z_2 gauge theory implying universality of this phenomenon across a wide range of theories with global and gauge symmetries.

C   Numerical simulations of field theories

C.1   Lattice simulations of Ising model and random field Ising model (RFIM)

The Ising model approximates (Euclidean) ϕ^4 quantum field theory in the continuum limit (achieved for RFIM at zero temperature [63]). Lattice simulations of zero-temperature RFIM in $D = 4, 5$ and 6 dimensions were performed on hypercubic lattices with sizes $L = 8, 10, 12, 16$ and 20. Ground states of the resulting IMs were calculated for 10^6 realizations of disorder. For both IM and RFIM, finite-size scaling effects were taken into account. After extraction of L-dependence, the values of critical exponents were determined by extrapolating $L^{-1} \to 0$. We obtained $\eta = 0.1942 \pm 0.0022$, $\nu = 0.8726 \pm 0.0182$ for $D = 4$, $\eta = 0.0442 \pm 0.0032$, $\nu = 0.6293 \pm 0.0030$ for $D = 5$ and $\eta = 0.0103 \pm 0.0041$, $\nu = 0.4892 \pm 0.0171$ for $D = 6$.

C.2   Lattice simulations of pure and random field Z_2 gauge field theories

Monte-Carlo lattice simulations of Euclidean Z_2 and RF (random field) Z_2 gauge field theories were performed on periodic hypercubic lattices of the size $L = 8, 12, 18, 24$ and 28 for the spatial part and fixed $L = 2$ for the inverse temperature part of the lattice. For the RF Z_2 gauge field theory, 10^6 realizations of random disorder were used. For Z_2 gauge field theory, we obtained $\beta = 0.13 \pm 0.02$, $\nu = 0.99 \pm 0.03$ for $D = 2 + 1$ and $\beta = 0.33 \pm 0.01$, $\nu = 0.63 \pm 0.03$ for $D = 3 + 1$ dimensions. (As usual, $+1$ denotes a dimension with periodic Matsubara boundary conditions.) For RF Z_2 gauge field theory, we found $\beta = 0.11 \pm 0.03$, $\nu = 0.65 \pm 0.04$ for $D = 4 + 1$ and $\beta = 0.30 \pm 0.05$, $\nu = 0.65 \pm 0.04$ for $D = 5 + 1$ dimensions.

D   Deriving Starobinsky-Fokker-Planck equation using path integral approach

In this appendix we shall illustrate the power of path integral approach for analyzing infrared dynamics of the scalar field in quasi-de Sitter universe by deriving the standard inflationary Fokker-Planck equation in the one-noise model. As usual we start with

$$\frac{\partial \Phi}{\partial N} = -\frac{1}{3H^2}\frac{\partial V}{\partial \Phi} + \frac{f}{H},$$ (D.1)

where the noise $f = H\sigma$ possesses the correlation properties

$$\langle \sigma(N)\sigma(N') \rangle = \frac{H^2}{4\pi^2}\delta(N - N').$$ (D.2)

This equation is derived straightforwardly using the approach described in [38] under the assumption of vanishing slow roll parameters ϵ_H, $\eta_H \to 0$. The partition function of the effective IR theory thus has the form

$$Z = \int \mathcal{D}\Phi \mathcal{D}\sigma \delta\left(\frac{\partial \Phi}{\partial N} + \frac{1}{3H^2}\frac{\partial V}{\partial \Phi} - \sigma\right)\exp\left(-\int dN \frac{2\pi^2}{H^2}\sigma^2\right).$$

Introducing a Lagrangian multiplier for the functional delta function, integrating out the noise σ as well as the Langrangian multiplier, we obtain

$$Z = \int \mathcal{D}\Phi \exp\left(-\int dN \frac{2\pi^2}{H^2}\left(\frac{\partial \Phi}{\partial N} + \frac{1}{3H^2}\frac{\partial V}{\partial \Phi}\right)^2\right).$$ (D.3)

Thus the Lagrangian of the theory is

$$\mathcal{L} = \frac{2\pi^2}{H^2}\left(\frac{\partial \Phi}{\partial N} + \frac{1}{3H^2}\frac{\partial V}{\partial \Phi}\right)^2.$$ (D.4)

The momentum conjugate to Φ is given by

$$P_\Phi = \frac{\partial \mathcal{L}}{\partial \Phi'} = \frac{\pi^2}{H^2}\left(\frac{\partial \Phi}{\partial N} + \frac{1}{3H^2}\frac{\partial V}{\partial \Phi}\right),$$ (D.5)

and the Hamiltonian of the theory corresponding to the Lagrangian (D.4) is

$$H_\Phi = P_\Phi \Phi' - \mathcal{L} = P_\Phi \Phi' - \mathcal{L} = \frac{1}{8\pi^2}P_\Phi H^2 P_\Phi - \frac{P_\Phi}{3H^2}\frac{\partial V}{\partial \Phi}.$$ (D.6)

The Starobinsky-Fokker-Planck equation [38] describing IR inflationary dynamics is obtained using this Hamiltonian and replacing $P_\Phi \to -\frac{\partial}{\partial \Phi}$ in the same fashion as Schroedinger equation is derived from the Feynman path integral for quantum mechanics:

$$\frac{\partial P(\Phi, N)}{\partial N} = H_\Phi\left(-\frac{\partial}{\partial \Phi}, \Phi\right)P(\Phi, N).$$ (D.7)

E   Non-locality in the one-noise model

E.1   Useful preliminary expressions and used notations

E.1.1   Slow roll parameters

In what follows, we consider the case of a single scalar field with a potential $V(\phi)$ propagating in a FRW spacetime with metric $ds^2 = dt^2 - a^2(t)d\mathbf{x}^2 = a^2(t)(d\eta^2 - d\mathbf{x}^2)$. Ignoring dependence on spatial coordinates $\mathbf{x}$, slow roll parameters are defined according to the usual prescription

$$\epsilon_H = \frac{M_P^2}{4\pi}\left(\frac{dH/d\phi}{H}\right)^2, \tag{E.1}$$

$$\eta_H = \frac{M_P^2}{4\pi}\frac{d^2H/d\phi^2}{H}. \tag{E.2}$$

Using the Hamilton-Jacobi equation for inflation

$$\left(\frac{dH}{d\phi}\right)^2 - \frac{12}{M_P^2}H^2 = \frac{32\pi^2}{M_P^4}V(\phi) \tag{E.3}$$

and expressions

$$\frac{d\phi}{dt} = -\frac{M_P^2}{4\pi}\frac{dH}{d\phi}, \quad \frac{dH}{dt} = -\frac{4\pi}{M_P^2}\left(\frac{d\phi}{dt}\right)^2 \tag{E.4}$$

one can demonstrate that

$$\epsilon_H = -\frac{dH/dt}{H^2}. \tag{E.5}$$

Using slow roll parameters ϵ_H, η_H rather than the usual slow roll parameters ϵ_V, η_V is more convenient as the end of inflationary stage corresponds to the condition $\epsilon_H = 1$ being held exactly.

Other useful formulae which we use below include

$$\frac{\partial(aH)}{\partial\eta} = (aH)^2(1 - \epsilon_H),$$

$$\frac{\partial\epsilon_H}{\partial\eta} = -2(\epsilon_H - \eta_H)\epsilon_H aH, \tag{E.6}$$

$$\frac{\partial^2(aH)}{\partial\eta^2} = (1 - 2\epsilon_H + 2\epsilon_H^2 - \epsilon_H\eta_H)(aH)^3.$$

The formula (E.6) rewritten in terms of inflationary efoldings $dN = Had\eta$ shows that taking time derivatives of slow roll parameters produces terms of higher order in slow roll expansion.

E.1.2   Number of inflationary efoldings

The Langevin and Fokker-Planck equations derived below are written in terms of the number of efoldings $N = \log a$ rather than the world time t or conformal time η. It is therefore appropriate to introduce the Jacobians associated with the corresponding change of variables. We find:

$$\frac{\partial}{\partial\eta} = aH\frac{\partial}{\partial N}.$$

A number of useful formulae which will be used in later derivations follow

$$\epsilon_H = -\frac{1}{H}\frac{dH}{dN}, \quad \eta_H = \epsilon_H - \frac{M_P}{\sqrt{16\pi}}\frac{d\epsilon_H/d\phi}{\sqrt{\epsilon_H}},$$

$$\frac{a''}{a} = (2 - \epsilon_H)(Ha)^2,$$

$$\frac{\partial}{\partial N}(aH) = aH(1 - \epsilon_H).$$

E.2   Separating scalar field into IR and UV parts. Langevin equation

The separation of the field into subhorizon and superhorizon parts is done according to

$$\phi = \Phi + \frac{1}{(2\pi)^{3/2}}\int d^3k\,\theta(k - \delta aH)\left(a_k\phi_k(\eta)e^{-i\mathbf{kx}} + \text{h.c.}\right). \tag{E.7}$$

Here as usual $\theta(\ldots)$ is the Heaviside step function of the argument and δ is a free dimensionless parameter identified with IR/UV separation scale (usually, in stochastic formalism it is taken to be small, $\delta \ll 1$, but not too small in order for potential terms to remain sub-dominant). Substituting the expression into the operator equation $\Box\phi + \frac{\partial V}{\partial\phi} = 0$ and neglecting potential term for the UV part of the field, we obtain:

$$\Box\Phi + \frac{1}{a^3}(u'' - \nabla^2 u - \frac{a''}{a}u + \ldots) \approx -\frac{\partial V}{\partial\Phi}, \tag{E.8}$$

where $\ldots$ denotes terms related to the potential and

$$\Box\Phi = H^2\frac{\partial^2\Phi}{\partial N^2} + H^2(3 - \epsilon_H)\frac{\partial\Phi}{\partial N} - \frac{\nabla^2\Phi}{a^2}$$

and

$$u = \frac{a}{(2\pi)^{3/2}}\int d^3k\,\theta(k - \delta aH)\left(a_k u_k(\eta)e^{-i\mathbf{kx}} + \text{h.c.}\right).$$

Substituting u into the eq. (E.8) and using the expression

$$u'' = -\frac{\delta^2}{(2\pi)^{3/2}}\int d^3k\,\delta'(k - \delta aH)(aH)^4(1 - \epsilon_H)^2(a_k u_k e^{-i\mathbf{kx}} + \text{h.c.})-$$

$$-\frac{2\delta}{(2\pi)^{3/2}}\int d^3k\,\delta(k - \delta aH)(aH)^3(1 - 2\epsilon_H + 2\epsilon_H^2 - \epsilon_H\eta_H)(a_k u_k e^{-i\mathbf{kx}} + \text{h.c.})-$$

$$-\frac{2\delta}{(2\pi)^{3/2}}\int d^3k\,\delta(k - \delta aH)(aH)^2(1 - \epsilon_H)(a_k u_k' e^{-i\mathbf{kx}} + \text{h.c.})+$$

$$+\frac{1}{(2\pi)^{3/2}}\int d^3k\,\theta(k - \delta aH)(a_k u_k'' e^{-i\mathbf{kx}} + \text{h.c.}),$$

we finally obtain the "Langevin" equation for the infrared part of the field

$$\frac{\partial^2\Phi}{\partial N^2} + (3 - \epsilon_H)\frac{\partial\Phi}{N} - \frac{\nabla^2\Phi}{(aH)^2} + \frac{\partial^2 V}{\partial\Phi^2} = f_1(t,\mathbf{x}) + f_2(t,\mathbf{x}), \tag{E.9}$$

where the "noise" operators f_1 and f_2 are defined according to

$$f_1 = -\frac{\delta^2 aH^2(1 - \epsilon_H)^2}{(2\pi)^{3/2}}\int d^3k\,\delta'(k - \delta aH)(a_k u_k e^{-i\mathbf{kx}} + \text{h.c.}), \tag{E.10}$$

$$f_2 = -\frac{2\delta}{(2\pi)^{3/2}}\int d^3k\,\delta(k - \delta aH)\left[H(1 - 2\epsilon_H + 2\epsilon_H^2 - \epsilon_H\eta_H)(a_k u_k e^{-i\mathbf{kx}} + \text{h.c.})+\right.$$

$$\left.\frac{1 - \epsilon_H}{a}(a_k u_k' e^{-i\mathbf{kx}} + \text{h.c.})\right]. \tag{E.11}$$

The part of the noise term f_1 seems to be suppressed at small $\delta \ll 1$. However, as we shall see, generally this is not the case as powers of δ are canceled out in the observable quantities.

E.3 Commutation relations of the noise operators f_1 and f_2

While the noise terms commute with terms on the r.h.s. of the eq. (E.9), it is also useful to check their self-commutation relations. Due to ultra-locality in the quasi-de Sitter regime, we shall be particularly interested in commutation relations of operators at the same spatial point $\mathbf{x}$. We find that

$$[f_1(N), f_1(N')] = 0, \tag{E.12}$$
$$[f_2(N), f_2(N')] = 0, \tag{E.13}$$
$$[f_1(N), f_2(N')] = -\frac{i\delta^3 H^2(1 - \epsilon_H)}{\pi^2}\delta'(N - N'). \tag{E.14}$$

It is worth noting that although the Langevin equation (E.9) is considered to be quasi-classical, operators f_1 and f_2 are not generally commuting although their commutator is small at $\delta \ll 1$ and becomes vanishing by the end of inflation when $\epsilon_H \to 1$.

This is in contrast with the Langevin-Starobinsky equation [38] for a scalar field on a fixed de Sitter background

$$\dot{\Phi} + \frac{1}{3H_0}\frac{\partial V}{\partial \Phi} = f, \tag{E.15}$$

where

$$f = -\frac{\delta a H_0^3 i}{4\pi^{3/2}} \int d^3k \delta(k - \delta a H_0)\frac{1}{k^{3/2}} \left(a_k e^{-i\mathbf{kx}} - a_k^\dagger e^{i\mathbf{kx}}\right). \tag{E.16}$$

One can immediately see that due to antisymmetric form of the combination $\left(a_k e^{-i\mathbf{kx}} - a_k^\dagger e^{i\mathbf{kx}}\right)$ the noise term $f(t)$ commutes with itself if the same spatial point is considered:

$$[f(t, \mathbf{x}), f(t', \mathbf{x})] = 0.$$

For points with large spatial (superhorizon) separation the noise terms do not commute even in this simplified case:

$$[f(t, \mathbf{x}), f(t', \mathbf{x}')] = \frac{H^3}{2\pi^2}\sin(\epsilon a H|\mathbf{x} - \mathbf{x}'|)\delta(t - t').$$

The reason of this discrepancy with our result is due to dropping terms $\sim f_1$ as suppressed by additional powers of δ at $\delta \ll 1$ during the derivation of (E.16); on the other hand, when deriving (E.12)–(E.14) all terms are kept explicitly. It is thus useful to remember that the quantum nature of the noise in the Langevin equation (E.9) is not eradicated completely during the quasi-de Sitter inflationary stage when $\epsilon_H \ll 1$.

E.4 Correlation functions of the noise operators f_1 and f_2

While we consider the field operators $f_1(N, \mathbf{x})$ and $f_2(N, \mathbf{x})$ quasi-classical quantities (based on their commutation relations in the regime $\delta \ll 1$ as well as their commuting with other terms in the Langevin equation (E.9)) in what follows, it is necessary to determine their stochastic properties. Those are given by the expectation values in the vacuum state of the Fock space of modes u_k. We find:

$$\langle f_2(N,\mathbf{x})f_2(N',\mathbf{x})\rangle = \frac{2\delta^2 H}{\pi^2}\left[\left(\frac{1-2\epsilon_H+2\epsilon_H^2-\epsilon_H\eta_H}{\sqrt{1-\epsilon_H}}\mathrm{Re}u_k + \frac{\sqrt{1-\epsilon_H}}{Ha}\mathrm{Re}u'_k\right)^2 + \right.$$
$$\left. +\left(\frac{1-2\epsilon_H+2\epsilon_H^2-\epsilon_H\eta_H}{\sqrt{1-\epsilon_H}}\mathrm{Im}u_k + \frac{\sqrt{1-\epsilon_H}}{Ha}\mathrm{Im}u'_k\right)^2_{k=\delta aH}\right]\delta(N-N'),$$

$$(E.17)$$

$$\langle f_1(N,\mathbf{x})f_1(N',\mathbf{x})\rangle = -\frac{H^2\delta^2}{2\pi^2}\left[(1-\epsilon_H)\delta aH(u_k u_k^*)_{k=\delta aH}\delta''(N-N')+\right.$$
$$+2(1-\epsilon_H)^2\delta aH(u_k u_k^*)_{k=\delta aH}\delta'(N-N')+$$
$$\left.+(1-\epsilon_H)^2(\delta aH)^2\left(\frac{d}{dk}(u_k u_k^*)\right)_{k=\delta aH}\delta'(N-N')\right],$$

$$(E.18)$$

$$\langle f_1(N,\mathbf{x})f_2(N',\mathbf{x})\rangle = \frac{\delta^3 H^2}{\pi^2}\left[aH(1-2\epsilon_H+2\epsilon_H^2-\epsilon_H\eta_H)(u_k u_k^*)_{k=\delta aH}+\right.$$
$$\left.+(1-\epsilon_H)(u_k u_k^{*\prime})_{k=\delta aH}\right]\delta'(N-N').$$

$$(E.19)$$

We emphasize that the expressions (E.17)–(E.19) are exact to all orders in slow roll parameters ϵ_H, η_H. They can be significantly simplified if the leading order in slow roll parameters is kept; in the regime $\eta_H \to 0$, $\epsilon_H \to 0$ we have

$$u_k \approx \frac{1}{\sqrt{2k}}\left(\frac{i}{k\eta}-1\right)e^{-ik\eta} = -\frac{1}{\sqrt{2k}}\left(\frac{iHa}{k}+1\right)e^{\frac{ik}{Ha}},$$
$$u'_k \approx \frac{1}{\sqrt{2k\eta}}e^{-ik\eta}\left(-\frac{i}{k\eta}-ik\eta+1\right) = -\frac{Ha}{\sqrt{2k}}e^{\frac{ik}{Ha}}\left(\frac{iHa}{k}+\frac{ik}{Ha}+1\right),$$

and

$$\langle f_2(N,\mathbf{x})f_2(N',\mathbf{x})\rangle = \frac{4H^2}{\pi^2}\delta(N-N'),$$

$$(E.20)$$

$$\langle f_1(N,\mathbf{x})f_1(N',\mathbf{x})\rangle = \frac{H^2}{8\pi^2}\delta'(N-N') - \frac{H^2}{4\pi^2}\delta''(N-N'),$$

$$(E.21)$$

$$\langle f_1(N,\mathbf{x})f_2(N',\mathbf{x})\rangle = \frac{H^2}{\pi^2}\delta'(N-N').$$

$$(E.22)$$

We are thus forced to conclude that in the general case the one noise model produces *a non-local* effective theory, and a rather hard one to deal with. This can immediately be seen from the behavior of the correlation functions (E.20)–(E.22) as well as after integrating the noise $f_1 f_2$ out in the partition function of the theory. The non-locality cannot be really neglected as the correlation functions of the noise $\delta'(N-N')$, $\delta''(N-N')$ are not suppressed. It also hints on the presence of an additional stochastic degree of freedom which was integrated out to produce the non-local behavior and forces us to apply the two-noise model described in the main text.

E.4.1   Relation to the Starobinsky's stochastic formalism

Two observations are in order. First of all, we note that even keeping f_2 only (which is essentially equivalent to using the approximation $\epsilon_H \ll 1$, $\delta \ll 1$ employed in [38] and throughout the literature) we do not reproduce Starobinsky's result for the numerical factor in front of the correlation function (E.20). Namely, the difference between the two results is a factor of 3/4, which is crucial given that it determines the correct value for de Sitter entropy. To understand what happens, let us recall how it is derived. If all terms suppressed by higher powers of slow roll parameters are neglected, the resulting equation for the superhorizon part of the field has the form

$$\frac{\partial \Phi}{\partial t} = -\frac{1}{3H}\frac{\partial V}{\partial \Phi} + f,$$

where

$$f = -\frac{i\delta a H^3}{4\pi^{3/2}} \int d^3k \delta(k - \delta a H)\frac{1}{k^{3/2}} \left(a_k e^{-i\mathbf{kx}} - a_k^\dagger e^{-i\mathbf{kx}}\right).$$

Note however that the terms $\ddot{\phi}$ which we neglected in the equations above would also contain the contribution $\sim \dot{f}$. While most contributions to $\dot{f}$ are suppressed by additional powers of δ, there is also a contribution present which is proportional to $\sim Hf$. This contribution is exactly the one which accounts for the difference between our and Starobinsky's result. However, keeping terms like this, we should be extra careful since f is a stochastic variable, which we are trying to differentiate.

Second, we note that the correlators E.19 and E.18 are *not* suppressed by powers of δ and thus should generally be kept. The resulting theory (after the noise term $f1$ is integrated out) is non-local in N. This non-locality hints on an existence of an additional effective field variable which has been integrated out to obtain the resulting non-local theory.

Acknowledgments

The work of A.O.B. was supported by the RFBR grant No.20-02-00297 and by the Foundation for Theoretical Physics Development "Basis".

References

[1] A. Einstein, B. Podolsky and N. Rosen, *Can quantum mechanical description of physical reality be considered complete?*, Phys. Rev. **47** (1935) 777 [INSPIRE].

[2] G. 't Hooft and M.J.G. Veltman, *One loop divergencies in the theory of gravitation*, Ann. Inst. H. Poincare Phys. Theor. **A20** (1974) 69.

[3] S. Deser and P. van Nieuwenhuizen, *One Loop Divergences of Quantized Einstein-Maxwell Fields*, Phys. Rev. D **10** (1974) 401 [INSPIRE].

[4] S.N. Solodukhin, *Entanglement entropy of black holes*, Living Rev. Rel. **14** (2011) 8 [arXiv:1104.3712] [INSPIRE].

[5] S.W. Hawking, *Breakdown of Predictability in Gravitational Collapse*, Phys. Rev. D **14** (1976) 2460 [INSPIRE].

[6] S. Weinberg, *The Cosmological Constant Problem*, Rev. Mod. Phys. **61** (1989) 1 [INSPIRE].

[7] S. Weinberg, *Critical phenomena for field theorists*, in *Proceedings of the International School of Subnuclear Physics, Ettore Majorana Center for scientific culture*, Erice, Italy, 24–26 July 1976, pp. 1–52.

[8] H.W. Hamber, *Quantum Gravity on the Lattice*, Gen. Rel. Grav. **41** (2009) 817 [arXiv:0901.0964] [INSPIRE].

[9] J. Ambjørn and J. Jurkiewicz, *Four-dimensional simplicial quantum gravity*, Phys. Lett. B **278** (1992) 42 [INSPIRE].

[10] S. Catterall, J.B. Kogut and R. Renken, *Phase structure of four-dimensional simplicial quantum gravity*, Phys. Lett. B **328** (1994) 277 [hep-lat/9401026] [INSPIRE].

[11] P. Bialas, Z. Burda, A. Krzywicki and B. Petersson, *Focusing on the fixed point of 4 − D simplicial gravity*, Nucl. Phys. B **472** (1996) 293 [hep-lat/9601024] [INSPIRE].

[12] M. Reuter, *Nonperturbative evolution equation for quantum gravity*, Phys. Rev. D **57** (1998) 971 [hep-th/9605030] [INSPIRE].

[13] N. Arkani-Hamed, L. Motl, A. Nicolis and C. Vafa, *The String landscape, black holes and gravity as the weakest force*, JHEP **06** (2007) 060 [hep-th/0601001] [INSPIRE].

[14] A. Strominger and C. Vafa, *Microscopic origin of the Bekenstein-Hawking entropy*, Phys. Lett. B **379** (1996) 99 [hep-th/9601029] [INSPIRE].

[15] J. Kaplan and S. Kundu, *Closed Strings and Weak Gravity from Higher-Spin Causality*, JHEP **02** (2021) 145 [arXiv:2008.05477] [INSPIRE].

[16] G. Parisi and N. Sourlas, *Random Magnetic Fields, Supersymmetry and Negative Dimensions*, Phys. Rev. Lett. **43** (1979) 744 [INSPIRE].

[17] A.M. Polyakov, *Quantum Geometry of Bosonic Strings*, Phys. Lett. B **103** (1981) 207 [INSPIRE].

[18] V.G. Knizhnik, A.M. Polyakov and A.B. Zamolodchikov, *Fractal Structure of 2D Quantum Gravity*, Mod. Phys. Lett. A **3** (1988) 819 [INSPIRE].

[19] T. Regge, *General relativity without coordinates*, Nuovo Cim. **19** (1961) 558 [INSPIRE].

[20] J.A. Wheeler, *Geometrodynamics and the Issue of the Final State*, in *Relativity Groups and Topology*, Les Houches Lecture Notes (1963).

[21] H.W. Hamber, *On the gravitational scaling dimensions*, Phys. Rev. D **61** (2000) 124008 [hep-th/9912246] [INSPIRE].

[22] W.G. Unruh, *Notes on black hole evaporation*, Phys. Rev. D **14** (1976) 870 [INSPIRE].

[23] B. DeWitt, *Quantum gravity, the new synthesis*, in *General Relativity: An Einstein centenary survey*, S.W. Hawking and W. Israel eds., Cambridge University Press (1979), pp. 680–745.

[24] S. Weinberg, *Ultraviolet divergences in quantum theories of gravitation*, in *General Relativity: An Einstein centenary survey*, S.W. Hawking and W. Israel eds., Cambridge University Press (1979), pp. 790–832.

[25] H. Kawai and M. Ninomiya, *Renormalization Group and Quantum Gravity*, Nucl. Phys. B **336** (1990) 115 [INSPIRE].

[26] T. Aida, Y. Kitazawa, J. Nishimura and A. Tsuchiya, *Two loop renormalization in quantum gravity near two-dimensions*, Nucl. Phys. B **444** (1995) 353 [hep-th/9501056] [INSPIRE].

[27] H.W. Hamber and R.M. Williams, *Higher derivative quantum gravity on a simplicial lattice*, Nucl. Phys. B **248** (1984) 392 [*Erratum ibid.* **260** (1985) 747] [INSPIRE].

[28] H.W. Hamber and R.M. Williams, *Nonperturbative simplicial quantum gravity*, Phys. Lett. B **157** (1985) 368 [INSPIRE].

[29] H.W. Hamber and R.M. Williams, *Gauge invariance in simplicial gravity*, Nucl. Phys. B **487** (1997) 345 [hep-th/9607153] [INSPIRE].

[30] H.W. Hamber and R.M. Williams, *Non-perturbative gravity and the spin of the lattice graviton*, Phys. Rev. D **70** (2004) 124007 [hep-th/0407039] [INSPIRE].

[31] B. Berg, *Exploratory numerical study of discrete quantum gravity*, Phys. Rev. Lett. **55** (1985) 904 [INSPIRE].

[32] A.A. Starobinsky, *A New Type of Isotropic Cosmological Models Without Singularity*, Phys. Lett. B **91** (1980) 99 [INSPIRE].

[33] J.D. Barrow and S. Cotsakis, *Inflation and the Conformal Structure of Higher Order Gravity Theories*, Phys. Lett. B **214** (1988) 515 [INSPIRE].

[34] A. De Felice and S. Tsujikawa, *f(R) theories*, Living Rev. Rel. **13** (2010) 3 [arXiv:1002.4928] [INSPIRE].

[35] A.O. Barvinsky and A.Y. Kamenshchik, *Quantum scale of inflation and particle physics of the early universe*, Phys. Lett. B **332** (1994) 270 [gr-qc/9404062] [INSPIRE].

[36] A.O. Barvinsky, A.Y. Kamenshchik and A.A. Starobinsky, *Inflation scenario via the Standard Model Higgs boson and LHC*, JCAP **11** (2008) 021 [arXiv:0809.2104] [INSPIRE].

[37] F.L. Bezrukov and M. Shaposhnikov, *The Standard Model Higgs boson as the inflaton*, Phys. Lett. B **659** (2008) 703 [arXiv:0710.3755] [INSPIRE].

[38] A.A. Starobinsky, *Stochastic de sitter (inflationary) stage in the early universe*, in *Field theory, quantum gravity and strings*, Berlin, Germany: Springer (1988), pp. 107–126.

[39] M. Sasaki, Y. Nambu and K.-i. Nakao, *Classical Behavior of a Scalar Field in the Inflationary Universe*, Nucl. Phys. B **308** (1988) 868 [INSPIRE].

[40] Y. Nambu and M. Sasaki, *Stochastic Approach to Chaotic Inflation and the Distribution of Universes*, Phys. Lett. B **219** (1989) 240 [INSPIRE].

[41] A.A. Starobinsky and J. Yokoyama, *Equilibrium state of a selfinteracting scalar field in the de Sitter background*, Phys. Rev. D **50** (1994) 6357 [astro-ph/9407016] [INSPIRE].

[42] S.-J. Rey, *Dynamics of Inflationary Phase Transition*, Nucl. Phys. B **284** (1987) 706 [INSPIRE].

[43] A. Hosoya, M. Morikawa and K. Nakayama, *Stochastic Dynamics of Scalar Field in the Inflationary Universe*, Int. J. Mod. Phys. A **4** (1989) 2613 [INSPIRE].

[44] F.R. Graziani, *Quantum Probability Distributions in the Early Universe. 1. Equilibrium Properties of the Wigner Equation*, Phys. Rev. D **38** (1988) 1122 [INSPIRE].

[45] I.D. Lawrie, *Perturbative Description of Dissipation in Nonequilibrium Field Theory*, Phys. Rev. D **40** (1989) 3330 [INSPIRE].

[46] V. Castellani, S. Degl'Innocenti, P.G. Prada Moroni and V. Tordiglione, *Hipparcos open clusters and stellar evolution*, Mon. Not. Roy. Astron. Soc. **334** (2002) 193 [astro-ph/0203327] [INSPIRE].

[47] K. Enqvist, S. Nurmi, D. Podolsky and G.I. Rigopoulos, *On the divergences of inflationary superhorizon perturbations*, JCAP **04** (2008) 025 [arXiv:0802.0395] [INSPIRE].

[48] D.I. Podolsky, *On Triviality of $\lambda\phi^4$ Quantum Field Theory in Four Dimensions*, arXiv:1003.3670 [INSPIRE].

[49] D. Podolskiy and R. Lanza, *On decoherence in quantum gravity*, Annalen Phys. **528** (2016) 663 [arXiv:1508.05377] [INSPIRE].

[50] D. Podolskiy, *Microscopic origin of de Sitter entropy*, arXiv:1801.03012 [INSPIRE].

[51] C. Pattison, V. Vennin, H. Assadullahi and D. Wands, *Stochastic inflation beyond slow roll*, JCAP **07** (2019) 031 [arXiv:1905.06300] [INSPIRE].

[52] A.M. Polyakov, *Infrared instability of the de Sitter space*, arXiv:1209.4135 [INSPIRE].

[53] C. Itzykson and J.M. Drouffe, *Statistical Field Theory*. Cambridge University Press, Cambridge, U.K. (1989).

[54] V.V. Slezov, *Kinetics of First-Order Phase Transitions*, Wiley-VCH Verlag GmbH & Co. KGaA, Weigheim, Germany (2009).

[55] M. Aizenman, *Geometric Analysis of ϕ^4 Fields and Ising Models (Parts 1 & 2)*, *Commun. Math. Phys.* **86** (1982) 1 [INSPIRE].

[56] M. Aizenman, *On the renormalized coupling constant and the susceptibility in ϕ^4 in four-dimensions field theory and the Ising model in four-dimensions*, *Nucl. Phys.* B **225** (1983) 261 [INSPIRE].

[57] W.H. Zurek, *Pointer Basis of Quantum Apparatus: Into What Mixture Does the Wave Packet Collapse?*, *Phys. Rev.* D **24** (1981) 1516 [INSPIRE].

[58] W.H. Zurek, *Decoherence, einselection, and the quantum origins of the classical*, *Rev. Mod. Phys.* **75** (2003) 715 [quant-ph/0105127] [INSPIRE].

[59] N.G. Fytas, V. Martín-Mayor, M. Picco and N. Sourlas, *Specific-heat exponent and modified hyperscaling in the 4D random-field Ising model*, *J. Stat. Mech.* **1703** (2017) 033302 [arXiv:1611.09015] [INSPIRE].

[60] N.G. Fytas and V. Martín-Mayor, *Universality in the Three-Dimensional Random-Field Ising Model*, *Phys. Rev. Lett.* **110** (2013) 227201 [*Erratum ibid.* **111** (2013) 019903; *Erratum ibid.* **111** (2013) 119903].

[61] N.G. Fytas, V. Martín-Mayor, M. Picco and N. Sourlas, *Phase Transitions in Disordered Systems: The Example of the Random-Field Ising Model in Four Dimensions*, *Phys. Rev. Lett.* **116** (2016) 227201.

[62] N.G. Fytas, V. Martin-Mayor, M. Picco and N. Sourlas, *Restoration of Dimensional Reduction in the Random-Field Ising Model at Five Dimensions*, *Phys. Rev.* E **95** (2017) 042117 [arXiv:1612.06156] [INSPIRE].

[63] D.S. Fisher, *Scaling and critical slowing down in random-field Ising systems*, *Phys. Rev. Lett.* **56** (1986) 416 [INSPIRE].

[64] J.C. Angles d'Auriac, M. Preissmann and R. Rammal, *The random field Ising model: algorithmic complexity and phase transition*, *J. Phys. Lett.* **46** (1985) 173.

[65] A. Goldberg and R. Tarjan, *Solving minimum-cost flow problems by successive approximation*, in *Proceedings of the nineteenth annual ACM conference on Theory of computing — STOC '87*, New York, U.S.A., 1987, pp. 7–18.

[66] R. Balian, J.M. Drouffe and C. Itzykson, *Gauge Fields on a Lattice. 2. Gauge Invariant Ising Model*, *Phys. Rev.* D **11** (1975) 2098 [INSPIRE].

[67] M. Creutz, *Phase Diagrams for Coupled Spin Gauge Systems*, *Phys. Rev.* D **21** (1980) 1006 [INSPIRE].

[68] E. Kehl, H. Satz and B. Waltl, *Critical exponents of Z_2 gauge theory in (3+1)-dimensions*, *Nucl. Phys.* B **305** (1988) 324 [INSPIRE].

FURTHER READING*

R. Lanza, with B. Berman, *Biocentrism* (BenBella Books, 2009) [Introduction].

R. Lanza, with B. Berman, *Beyond Biocentrism* (BenBella Books, 2016) [Introduction].

M. Pavšič, *The Landscape of Theoretical Physics: A Global View* (Kluwer Academic, 2001) [2, 7, 10].

H. Everett III, "The Theory of the Universal Wave Function," in *The Many Worlds Interpretation of Quantum Mechanics*, B. S. DeWitt, and N. Graham, editors (Princeton University Press, 1973) [2, 8, 10, 12].

M. Tribus and E. C. McIrvine, "Energy and Information," *Scientific American*, 224, 179 (1971) [4].

H. Stapp, "Quantum Approaches to Consciousness," in *The Cambridge Handbook of Consciousness*, edited by P. D. Zelazo (Cambridge University Press, 2007) [7].

---

* The numbers in the brackets denote chapters in this book that a reference pertains to.

R. Hofstadter, *Gödel, Escher, Bach: An Eternal Golden Braid* (Penguin Books, 1980) [7].

H. Zwirn, "The Measurement Problem: Decoherence and Convivial Solipsism," *Foundations of Physics* 46, 635 (2016) [7].

B. Libet, "Time of Conscious Intention to Act in Relation to Onset of Cerebral Activity" (Readiness-Potential), *Brain* 106, 623 (1983) [8].

D. Deutsch, "Quantum Mechanics Near Closed Timelike Lines," *Physical Review D*, 44, 3197 (1991) [8, 12].

M. Tegmark, "The Multiverse Hierarchy," in *Universe or Multiverse?*, B. Carr, editor (Cambridge University Press, 2007) [10].

R. P. Feynman, "Mathematical formulation of the quantum theory of electromagnetic interaction," *Physical Review*, 80, 440 (1950) [11].

E. C. G. Stueckelberg, *Helvetica Physica Acta*, 14, 322 (1941) [11, 12].

S. S. Schweber, "Feynman and the visualization of spacetime process," *Reviews of Modern Physics* 58, 449–505 (1986) [11, 12].

L. P. Horwitz and F. Rohrlich, *Physical Review D*, 30, 1528 (1981) [11, 12].

J. R. Fanchi, *Parametrized Relativistic Quantum Theory* (Kluwer Academic, 1993) [11, 12].

E. A. B. Cole, "Particle Decay in Six-Dimensional Relativity," *Journal of Physics A*, 13, 109 (1980) [11, 12].

M. Pavšič, "On the Quantization of Gravity by Embedding Spacetime in a Higher Dimensional Space," *Classical and Quantum Gravity*, 2, 869 (1985) [12, 15].

ACKNOWLEDGMENTS

The authors would like to thank the publisher, Glenn Yeffeth, as well as Alexa Stevenson and Pate Steele for their excellent editorial assistance. We would also like to thank Jacqueline Rogers for illustrations throughout the book, and Dmitriy Podolskiy for his help with Chapters 11 and 14. Various portions of the material in this book appeared separately in the *Huffington Post*, *Omni*, *Discover* magazine, and *Psychology Today*.

ABOUT THE AUTHORS

Robert Lanza

Robert Lanza, MD is one of the most respected scientists in the world—a *U.S. News & World Report* cover story called him a "genius" and "renegade thinker," even likening him to Einstein. Lanza is head of Astellas Global Regenerative Medicine, Chief Scientific Officer of the Astellas Institute for Regenerative Medicine, and adjunct professor at Wake Forest University School of Medicine. He was recognized by *TIME* magazine in 2014 on its list of the "100 Most Influential People in the World." *Prospect* magazine named him one of the Top 50 "World Thinkers" in 2015. He is credited with several hundred publications and inventions, and more than thirty scientific books, including the definitive references in the field of stem cells and regenerative medicine. A former Fulbright Scholar, he studied with polio pioneer Jonas Salk and Nobel Laureates Gerald Edelman and Rodney Porter. He also worked closely (and coauthored a series of papers) with noted Harvard psychologist B. F. Skinner and heart transplant pioneer Christiaan Barnard. Dr. Lanza received his undergraduate and medical degrees from the University of Pennsylvania, where he was both a University Scholar and Benjamin Franklin Scholar. Lanza was part of the team that cloned the world's first human embryo, as well as the first to successfully generate stem cells from adults using somatic-cell nuclear transfer (therapeutic cloning). In 2001 he was also the first to clone an endangered species, and recently published the first-ever report of pluripotent stem cell use in humans.

Matej Pavšič

Matej Pavšič is a physicist interested in foundations of theoretical physics. During his more than forty years of research at the Jožef Stefan Institute in Ljubljana, Slovenia, he often investigated the subjects that currently were not of wide interest, but later became hot topics. For example, in the 70s he studied higher dimensional, Kaluza-Klein, theories when they were not very popular, and in the 80s he proposed an early version of the braneworld scenario that was published, among others, in *Classical and Quantum Gravity*. Altogether, Matej Pavšič published more than hundred scientific papers and the book *The Landscape of Theoretical Physics: A Global View* (Kluwer Academic, 2001). He is among the pioneering authors in topics such as mirror particles, braneworld, Clifford space, and has recently published important works explaining why negative energies in higher derivative theories are not problematic, which is crucial for quantum gravity. Matej Pavšič studied physics at the University of Ljubljana. After obtaining his master degree in 1975, he spent a year at the Institute of Theoretical Physics in Catania, Italy, where he collaborated with Erasmo Recami and Piero Caldirola. Under their supervision he completed his PhD thesis which he later defended at the University of Ljubljana. Matej Pavšič participated at many conferences as an invited speaker and regularly visited the International Centre for Theoretical Physics (ICTP) in Trieste, where he worked with the famous theoretical physicist Asim O. Barut.

Bob Berman

Bob Berman is the longtime science editor of *The Old Farmer's Almanac* and contributing editor of *Astronomy* magazine, formerly with *Discover* from 1989 to 2006. He produces and narrates the weekly *Strange Universe* segment on WAMC Northeast Public Radio, heard in eight states, and has been a guest on such TV shows as *Late Night with David Letterman*. He taught physics and astronomy at New York's Marymount College in the 1990s and is the author of eight popular books. His newest is *Zoom: How Everything Moves* (2014, Little, Brown).

INDEX